FM 23-32

DEPARTMENT OF THE ARMY FIELD MANUAL

3.5-INCH
ROCKET LAUNCHER
"SUPER BAZOOKA"

FIELD MANUAL

By DEPARTMENT OF THE ARMY
DECEMBER 1961

©2013 Periscope Film LLC
All Rights Reserved
ISBN#978-1-940453-02-6
www.PeriscopeFilm.com

FIELD MANUAL

No. 23-32

HEADQUARTERS,
DEPARTMENT OF THE ARMY
WASHINGTON 25, D.C., *27 December 1961*

3.5-INCH ROCKET LAUNCHER

*This manual supersedes FM 23-32, 16 April 1958.

CHAPTER 1

INTRODUCTION

Section I. GENERAL

1. Purpose and Scope

a. This manual describes the characteristics, functioning, maintenance, steps of preparatory marksmanship, technique of fire, and range control for the standard 3.5-inch rocket launchers, M20A1 and M20A1B1 (figs. 1 and 2). It discusses the types of ammunition, including general characteristics, capabilities and limitations, and the functioning of the rocket (fig. 3). When range is given and refers to the sight or its usage, it is given in yards rather than converted to the metric system as prescribed in AR 700–75. This is done for the purpose of clarity, as the sight for the 3.5-inch rocket launcher is graduated in yards.

b. This manual is applicable to both nonnuclear and nuclear warfare.

c. Users of this manual are encouraged to submit recommended changes or comments to improve the manual. Comments should be keyed to the specific page, paragraph, and line of the text in which the change is recommended. Reasons should be provided for each comment to insure understanding and complete evaluation. Comments should be forwarded direct to the United States Army Infantry School, Fort Benning, Ga.

d. Appendix III contains the text material for launchers M20 and M20B1.

2. General Characteristics

a. Rocket Launcher. The rocket launcher is an extremely lightweight weapon due to a special aluminum alloy used in its construction. It provides close-in antitank protection by igniting a rocket and guiding it on its initial flight toward the target.

b. Rocket. The rocket consists of a warhead, fuze, and rocket motor. It is propelled by the reaction from a discharging jet of gas produced by the burning of a propellant charge inside the rocket motor. The rocket's momentum carries it to the target.

3. How Rocket Operates From Rocket Launcher

a. When a gas is compressed in a closed tube, the pressure in one direction is balanced by an equal pressure in the opposite direction

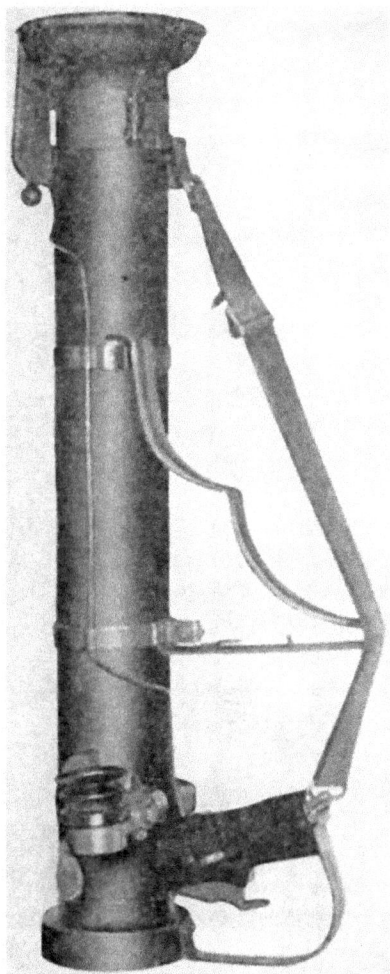

Figure 1. Rear barrel, 3.5-inch launcher, M20A1.

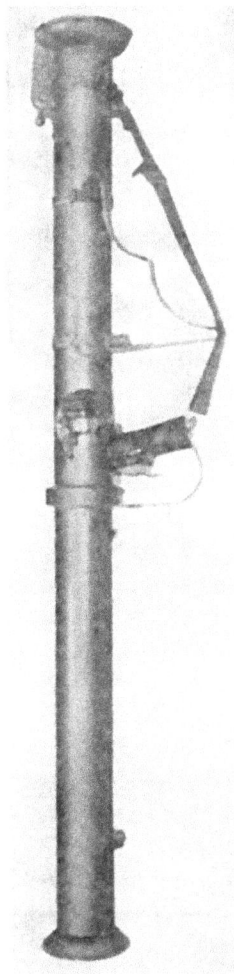

Figure 2. 3.5-inch rocket launcher, M20A1B1.

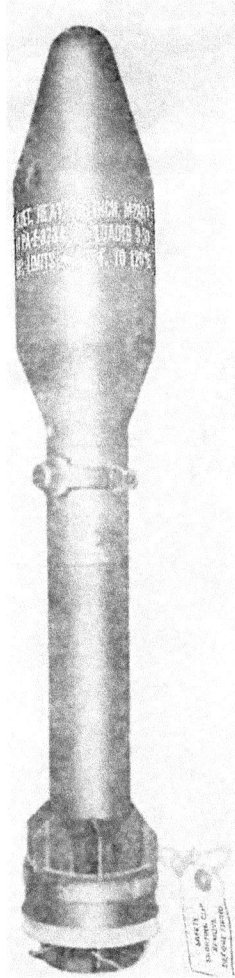

Figure 2. 3.5-inch rocket HEAT, M28A2.

(fig. 4). An opening at one end of the tube reduces the area against which the pressure is applied (fig. 5). When pressure is maintained inside the rocket motor tube, the total pressure against the closed (warhead) end is greater than the total pressure against the reduced area of the open (venturi) end. An unbalanced condition of pressure results, therefore, forcing the rocket forward and in the direction of the muzzle end of the launcher.

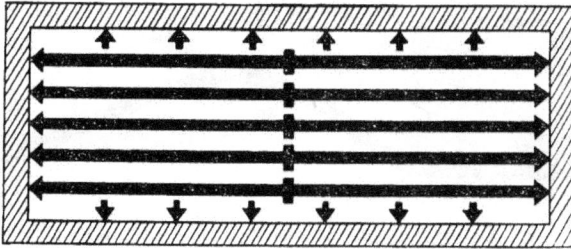

Figure 4. Closed tube.

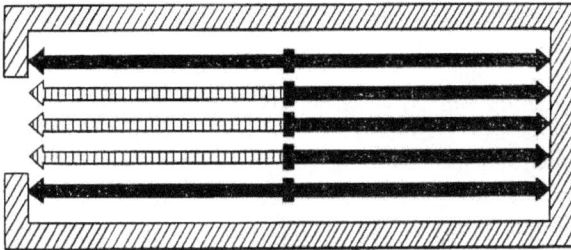

Figure 5. Open tube.

b. The rearward pressure in the rocket is reduced because the gases can escape through a nozzle opening. The forward pressure remains the same, and the rocket is forced forward and out of the launcher. The velocity and quantity of the gases escaping to the rear determine the velocity of the rocket.

c. The rocket motor, or stabilizer tube, is closed at the forward end and has a constricted opening at the rear called a nozzle or venturi (fig. 6). This opening makes the gas flow smoothly and steadily. It prevents the gas from escaping too fast and thus maintains the pressure within the rocket motor. The lateral expansion of the gas against the outer slope of the nozzle assists the rocket in its forward movement.

Section II. DESCRIPTION AND DATA

4. Description

a. The M20A1B1 launcher is a two-piece, smoothbore, open-tube weapon. When prepared for firing, the front and rear barrel groups

5

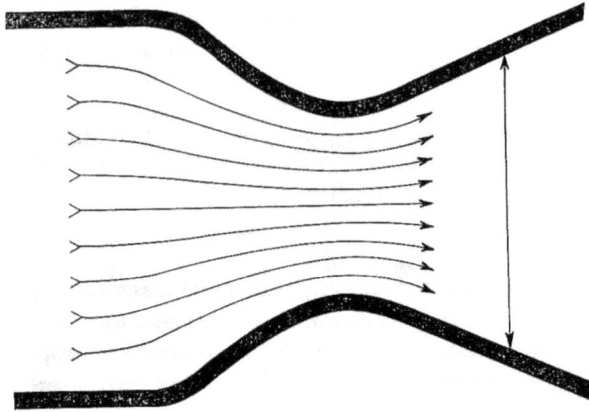

Figure 6.   A venturi tube.

are joined to form a launching tube (fig. 2). It is sighted on the target by a reflecting sight assembly. A magneto located in the trigger grips provides the current for firing the rocket. This weapon is fired from the shoulder in the standing, kneeling, modified kneeling, and prone positions.

b. A gun sling (fig. 2) is used to carry the launcher. In the carrying position, the barrels are fastened together in a double-tube arrangement, thus eliminating the unwieldy length of the assembled weapon. The front barrel of all models of the 3.5-inch rocket launcher is interchangeable; but the launcher must be boresighted after the barrel is changed.

c. The launcher ignites the propellant charge of the rocket and guides it on its initial stage of flight toward the target. The propulsion of the rocket is accomplished by the action of a burning propellant charge in the motor body of the rocket. The propulsion does not depend on gas pressure built up inside the launcher tube. The tube, therefore, is only heavy enough to prevent denting or bending during handling, and to prevent excessive heating at normal rates of fire.

d. When the propellant is ignited, gases and flames are blown from the breech of the launcher. The area directly to the rear of the launcher must be clear of personnel and materiel. Because of the blast to the rear of the weapon, the gunner must take special precautions to avoid injury to himself, especially when firing in the prone position.

5. Similarities Between Models

The M20A1 and M20A1B1 launchers have identical contactor latch group assemblies. The contactor latch group assembly on both models eliminates the necessity of hooking the contact wire of the rocket to the contact spring on the launcher as was required on earlier models. This permits an increased rate of fire. The models differ in that the component parts are cast on the front and rear barrels of the M20A1B1 launcher.

6. Tabulated Data

    a. Launchers.

Length of launcher assembled (for firing) (approx)_____ 60 inches
Weight of launcher M20A1B1_____ 13 pounds
Length of front barrel_____ 30 inches
Weight of front barrel, M20A1B1_____ 4 pounds
Length of rear barrel (approx)_____ 30 inches
Weight of rear barrel (approx)_____ 9 pounds
Type of firing mechanism_____ Electrical
Maximum effective range (point type targets)_____ 300 yards (275 M)
Maximum effective range (moving targets)_____ 200 yards (185 M)

    b. Ammunition.

Type used_____ HEAT (high explosive antitank) and TP (target practice)
Weight of rocket (approx)_____ 9 pounds
Maximum range (approx)_____ 900 yards (825 M)
Muzzle velocity at 70° F. (approx)_____ 334 feet-per-second

CHAPTER 2

LAUNCHER CONTROLS AND SIGHTING EQUIPMENT

Section I. CONTROLS

7. General

This section describes, locates, and illustrates the controls needed to operate the M20A1 and M20A1B1 rocket launchers.

8. Trigger

When the trigger is squeezed, a magnetic armature is placed in motion and an electrical firing mechanism (fig. 7) in the trigger grip is activated. This generates the electrical current used to ignite the rocket.

9. Safety Switch

The safety switch (fig. 7) on the left side of the trigger grip has two positions. In the lower (SAFE) position, a block on the safety mechanism prevents the trigger from moving. In the upper (FIRE) position, the block of the safety mechanism is held clear of a lug on the trigger, and the trigger is free to move.

Caution: Except when firing, the safety is always kept in the SAFE position.

Figure 7. Electrical firing mechanism.

10. Barrel Coupling Lock Lever

The barrel coupling lock lever (fig. 8) is on the right side of the rear barrel assembly just behind the coupling nut. It is used to release the barrel coupling nut. The barrel coupling nut and screw are so designed that a twist of less than one-quarter turn will engage or disengage them. The barrel coupling lock lever is operated by raising and holding the lever in its unlocked position. When the lever is released, a spring returns it to its locked position. It is used when disassembling and assembling the front and rear barrels.

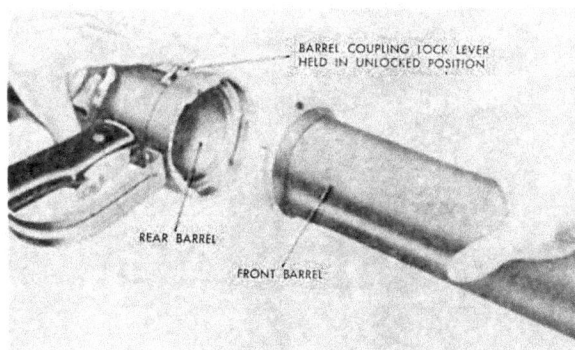

Figure 8.   Method of assembling barrels for firing.

11. Barrel Latch Handle

The barrel latch handle (fig. 9) is on the right side of the rear barrel assembly, just in rear of the trigger grip. It is used to release the barrel latch which locks the front and rear barrel assemblies in the carrying position. Moving the handle forward retracts the barrel latchbolt from the barrel latch strike on the front barrel assembly. Two springs on the barrel latchbolt keep the bolt in its normal extended position.

12. Contactor Latch Group Assembly, Models M20A1 and M20A1B1

The contactor latch group assembly (fig. 10) is mounted on the strap assembly in front of the breech guard (fig. 2). It consists of a control handle, right- and left-hand contact brackets (fig. 11), and a firing contact cable assembly (fig. 13). It also has a firing contact, a detent, right- and left-hand positioning stops, and a trip lever (fig. 12).

13. Contact Lead Cable and Contact Spring Clamp

The contact lead cable (fig. 13) is an insulated wire connecting the firing mechanism in the trigger grip to the contact spring clamp.

619881°—62——2

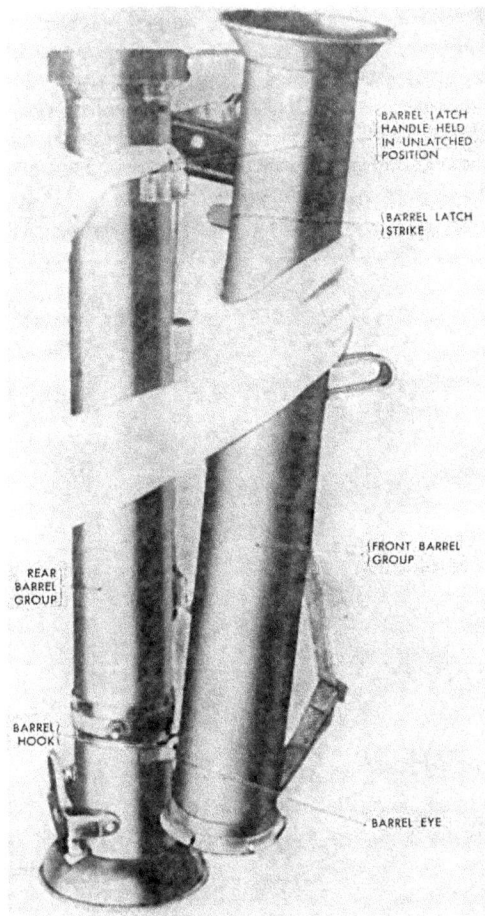

BARREL LATCH
HANDLE HELD
IN UNLATCHED
POSITION

BARREL LATCH
STRIKE

FRONT BARREL
GROUP

REAR
BARREL
GROUP

BARREL
HOOK

BARREL EYE

Figure 9. Unlatching barrels from carrying position.

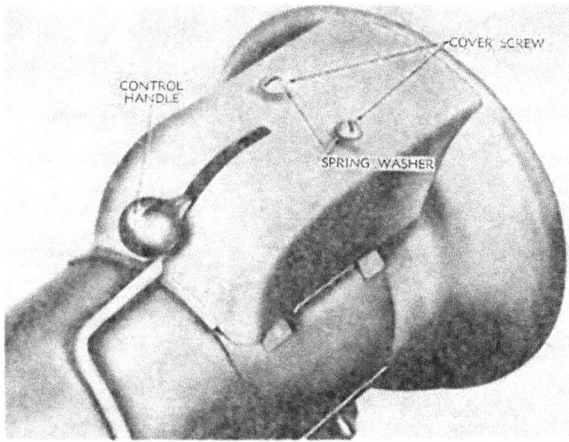

Figure 10. Contactor latch group assembly with control handle in LOAD position.

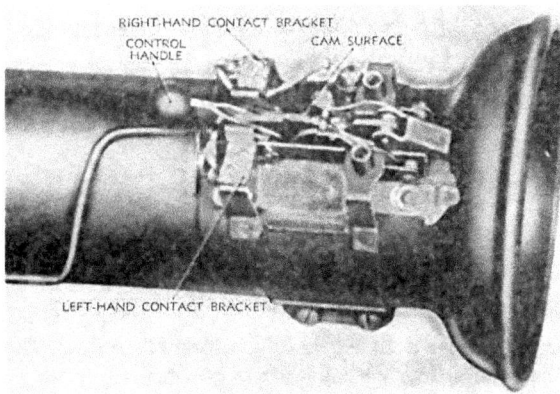

Figure 11. Control handle in LOAD position with cover removed.

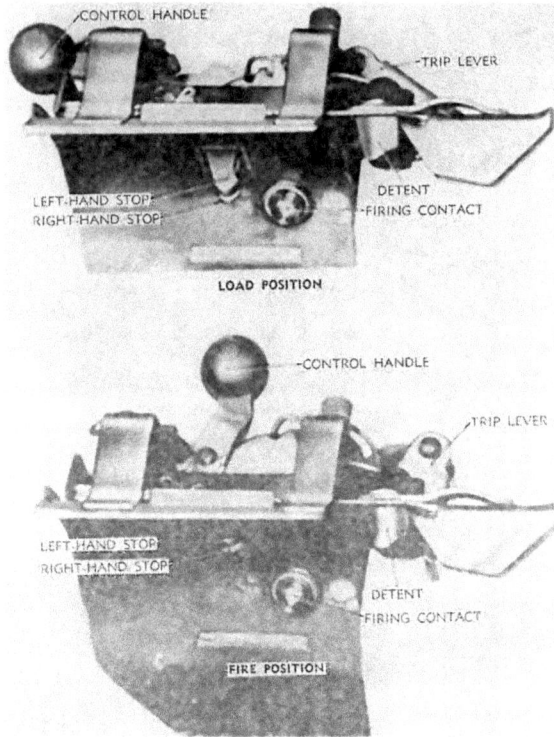

Figure 12. Contactor latch group assembly removed from launcher with positioning strap removed.

14. Breech Guard

The breech guard is a flared metal ring on the rear of the launcher. It facilitates the loading of rockets and protects the component parts of the contactor latch group assembly which protrude into the breech of the weapon. The breech guard also helps keep dirt out of the barrel when the rear of the launcher rests on the ground.

15. Stock

The stock is a piece of strap metal clamped around the rear barrel assembly and formed to fit the shoulder.

16. Sling

The sling is made of a standard 1½-inch webbing and has buckles for adjusting its length. One end of the sling is attached to the hand grip; the other is attached to a special bracket on the rear section of the barrel. The sling is used only to carry the launcher.

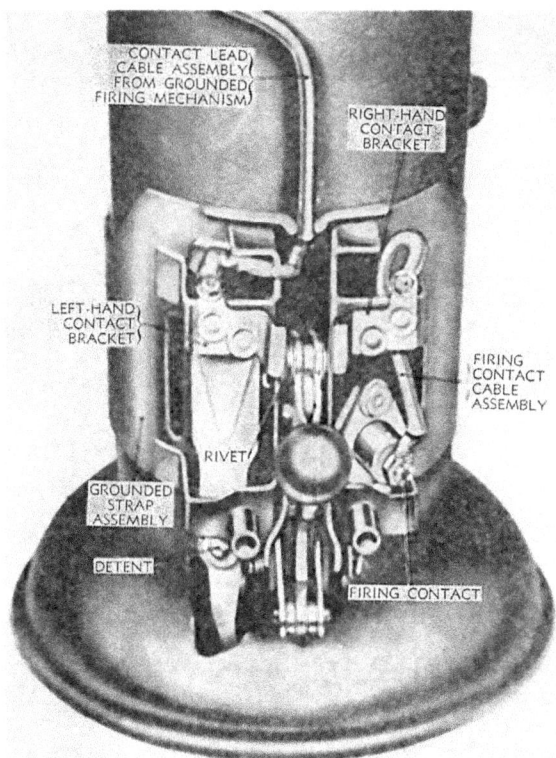

Figure 13. Contactor latch group assembly with cover removed.

17. Muzzle Flash Deflector

The muzzle flash deflector is a flared metal ring on the front end of the barrel. It protects the gunner and loader from being injured by particles of late-burning propellant and small particles of wire that might be blown back from the fin assembly of the rocket as it clears the muzzle of the weapon.

Section II. SIGHTING EQUIPMENT

18. General

Equipment used for sighting the M20A1 and M20A1B1 launchers consists of a reflecting sight assembly (fig. 14) and an elevation plate. These are mounted on the left side of the rear barrel assembly.

19. Reflecting Sight Assembly (Ladder or Tree-Type Reticle Pattern)

The reticle (fig. 15) in the reflecting sight assembly has a broken vertical center line, a broken horizontal zero line, and four broken

Figure 14. Reflecting sight assembly in firing position.

horizontal range lines. Each mark of the vertical center line and each space between marks represents 50 yards of range. Each mark of the horizontal line and each space between marks represents leads required for various target speeds. These progress to the right and left of the center line in 5-mph increments for a total of 30-mph of speed leads in either direction. The zero line is marked 0 at each end. The four range lines are each marked to represent the range in yards.

20. Range Scale

a. A range scale is engraved on the elevation plate (fig. 16). It is used in conjunction with the reflecting sight assembly (tree-type reticle pattern), and has a zero setting notch marked "0 to 450." It also has notches numbered from five to nine. A spring-loaded projection on the indicator arm pointer engages these notches, holding it at the desired range setting.

b. If the range is 450 yards or less, set the indicator arm at the "0 to 450" notch of the range scale and select the proper range on the sight reticle. If the range to the target is over 450 yards, move the indicator arm to the notch of the range scale corresponding to the desired range. Use the zero line of the sight reticle to obtain a sight picture. For example, if the range is estimated to be 600 yards (550 M), set the indicator arm at six on the range scale. Looking through the sight, set the zero line of the sight reticle on the target. If the target is stationary, use the vertical line; if the target is moving, take the appropriate lead.

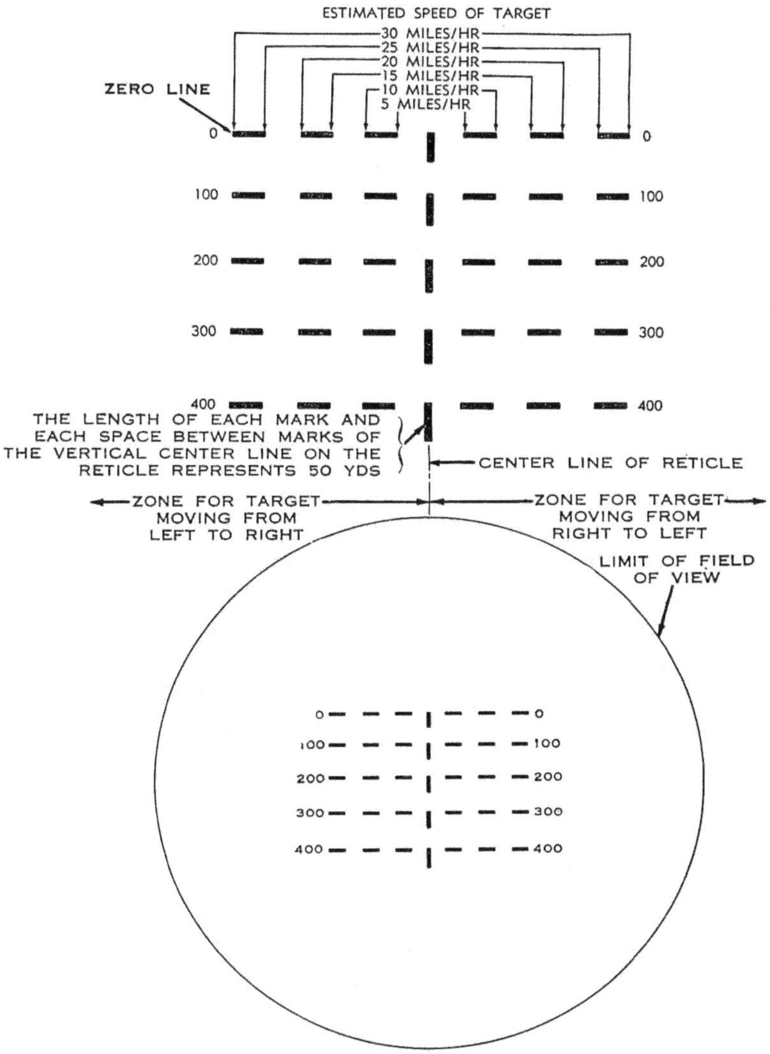

ESTIMATED SPEED OF TARGET

30 MILES/HR
25 MILES/HR
20 MILES/HR
15 MILES/HR
10 MILES/HR
5 MILES/HR

ZERO LINE

0 ▬ ▬ ▬ ▮ ▬ ▬ ▬ 0

100 ▬ ▬ ▬ ▮ ▬ ▬ ▬ 100

200 ▬ ▬ ▬ ▮ ▬ ▬ ▬ 200

300 ▬ ▬ ▬ ▮ ▬ ▬ ▬ 300

400 ▬ ▬ ▬ ▮ ▬ ▬ ▬ 400

THE LENGTH OF EACH MARK AND
EACH SPACE BETWEEN MARKS OF
THE VERTICAL CENTER LINE ON THE
RETICLE REPRESENTS 50 YDS

CENTER LINE OF RETICLE

◄── ZONE FOR TARGET ──
MOVING FROM
LEFT TO RIGHT

ZONE FOR TARGET ──►
MOVING FROM
RIGHT TO LEFT

LIMIT OF FIELD
OF VIEW

0 ▬ ▬ ▬ ▮ ▬ ▬ ▬ 0
100 ▬ ▬ ▬ ▮ ▬ ▬ ▬ 100
200 ▬ ▬ ▬ ▮ ▬ ▬ ▬ 200
300 ▬ ▬ ▬ ▮ ▬ ▬ ▬ 300
400 ▬ ▬ ▬ ▮ ▬ ▬ ▬ 400

Figure 15.   Ladder- or tree-type reticle pattern.

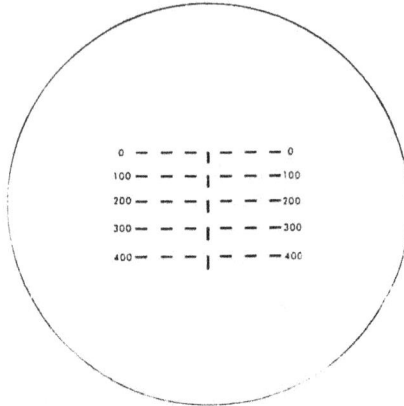

LADDER TYPE RETICLE PATTERN

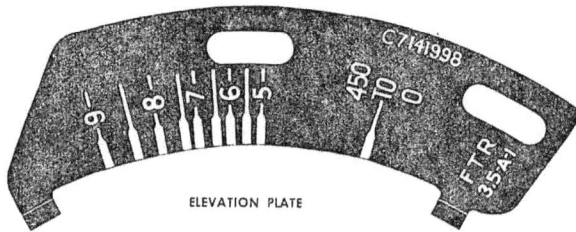

ELEVATION PLATE

Figure 16. Ladder-type reticle pattern and elevation plate.

21. Graduation on Indicator Arm Yoke

The reflecting sight lens frame (fig. 17) is secured to the indicator arm yoke by a hinge stud. This permits the sight to be folded back against the launcher barrel when not in use, or swung out into the firing position when needed. The head of the hinge stud (fig. 17) is engraved with an index set against an adjacent scale of five graduations on the indicator arm yoke. It is used for reference when setting the sight for deflection during boresighting.

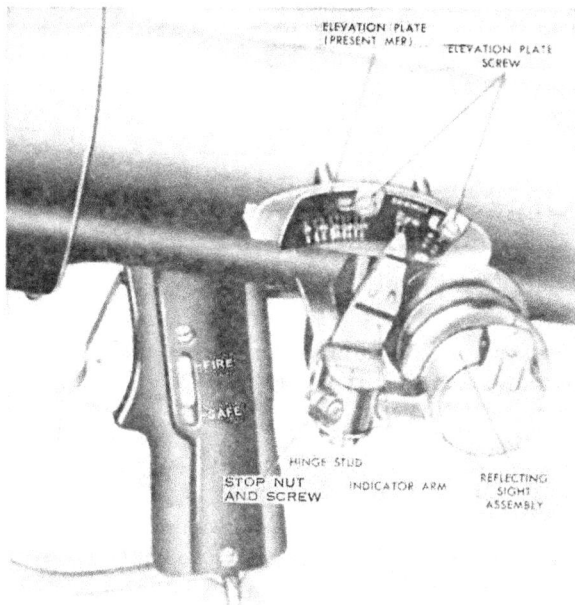

Figure 17. Reflecting sight assembly (in folded position) with elevation plate.

22. Manipulation of Reflecting Sight Assembly

a. The sight is unfolded from its carrying position by pulling it out and away from the barrel. This causes it to pivot on the sight hinge stud in the yoke of the indicator arm. It is swung out until the locking ball in the sight hinge stud engages and holds the sight in its sighting position.

b. After use, the sight is rotated back against the barrel until the locking ball snaps and secures the sight in its carrying position.

CHAPTER 3

OPERATION AND FUNCTIONING

23. Front and Rear Barrel Groups

a. *General.*

(1) The front and rear barrel groups are joined by the barrel coupling screw on the rear end of the front barrel and the barrel coupling nut on the forward end of the rear barrel (fig. 8). The barrel coupling screw and nut are so designed that a twist of less than one-quarter turn will engage or disengage them. The front barrel can be assembled to the rear barrel in any of three positions. If front barrels are interchanged between weapons, the weapons must be boresighted before use. The barrel coupling screw, nut, and both barrels are made of aluminum. On the M20A1B1 launcher these parts are cast as part of the barrels.

(2) The sight mounting bracket on the M20A1B1 launcher is a support for the sight and elevation plate. It is cast as a part of the rear barrel.

(3) The contact lead cable connects the contact latch group to the firing mechanism. The cable is encased in an aluminum tube to protect it against damage. The firing mechanism is grounded to the rear barrel.

b. *Electrical Firing Mechanism.* The electrical firing mechanism (fig. 7) is on the underside and the forward end of the rear barrel. It is secured to the barrel by means of a grip support and consists of the two grips, the trigger assembly, and the electrical firing mechanism. The electrical firing mechanism is composed of two magnets housing a coil of wire, which in turn houses a steel armature. When the trigger is squeezed, the armature rotates in the coil and generates sufficient current to ignite the rocket. This trigger mechanism generates current only when the trigger is squeezed. The trigger is shaped to fit the fingers. The electrical current generated by this trigger mechanism is almost three times that of earlier models, eliminating many misfires due to insufficient electrical current.

24. Preparing Launchers for Carrying and Firing

a. *Assembly.* To assemble the rocket launcher for firing—

(1) Raise the barrel latch handle and disengage the two barrels (fig. 9).

(2) Raise the barrel coupling lock lever and hold it in the unlocked position. Screw the barrel coupling screw of the front barrel into the barrel coupling nut of the rear barrel (fig. 8) and release the lock lever. Be sure the lock lever rotates into place. This locks the two barrels together.

(3) Unfold the sight and raise the lens cover. If the launcher is properly boresighted, it is ready for loading and firing.

b. *Disassembly.* To disassemble the launcher for carrying—

(1) Close the lens cover and rotate the sight back against the side of the barrel (fig. 17).

(2) Raise the barrel coupling lock lever (fig. 8) and unscrew the front barrel from the rear barrel.

(3) Place the front barrel hook into the rear barrel eye, and engage the barrel latch and barrel latch strike.

c. *Launchers, M20A1 and M20A1B1.* The contactor latch group assembly increases the rate of fire of the 3.5-inch rocket launcher by establishing electrical contact without the use of contact wires.

(1) The electrical circuit begins at the trigger mechanism and passes through the insulated contact lead cable to the left contact bracket (fig. 13). From the left contact bracket, it passes through the rivet (when in the firing position) to the right contact bracket, then through the firing contact cable to the firing contact. Continuing through the contact ring on the rocket igniter, it moves through the rocket to the detent, completing the circuit at the grounded strap assembly.

(2) With the control handle in LOAD position (fig. 10), the rivet in the control handle assembly does not make contact with either the right or left contact brackets (fig. 11), and the circuit is broken. The right and left positioning stops (fig. 12) protrude into the bore and the trip lever is in its UP position. The detent, the firing contact, and the positioning stops are spring-actuated. As the rocket is inserted into the barrel, the rocket head compresses the actuating springs. When the rocket head passes the detent, the firing contact, and the positioning stops, these parts move into the path of the fin assembly. The fin assembly is stopped by the positioning stops as the rocket is pushed forward. At this point, the firing contact engages the contact ring on the rocket. The detent engages the groove in the fin assembly, thereby positioning the rocket and retaining it in the tube.

(3) As the control handle is rotated to FIRE position, the positioning stops cam upward, clearing the bore. The trip lever (fig. 12) rotates downward into the rear of the bore. The rivet in the control handle assembly makes contact with both

contact brackets (fig. 13). In this position, the electrical circuit is complete.

(4) When the rocket is fired, the propellant gases strike the trip lever, forcing it upward and moving the control handle forward into the LOAD position. This action drops the positioning stops back into the bore, ready to receive another rocket.

CHAPTER 4
AMMUNITION

25. General Characteristics

a. Ammunition for the 3.5-inch rocket launcher is issued as a complete rocket of fixed ammunition (figs. 18 and 19). This ammunition can be fired from either the M20, M20A1, M20B1, or M20A1B1 rocket launcher models. The propelling charge is not adjustable, and the rocket is loaded into the launcher as a unit. The complete rocket consists of a rocket head, a fuze, and a rocket motor which contains the propellant and its igniter. A nozzle and fin assembly are rigidly attached to the rear of the motor. The fuze body, threaded at both ends, serves also as a coupling for the rocket head and motor.

b. Ammunition for the rocket launcher is classified according to the type of rocket head and includes high explosive antitank (HEAT) and practice (TP) rockets. Each rocket is about 23½ inches long, weighs about 9 pounds, has a maximum velocity of 334 feet-per-second at 70° F., and has an approximate maximum range of 900 yards (825 meters).

    (1) *High explosive antitank (HEAT) rocket.*

        (a) The head of the HEAT rocket, both the M28 and M28A2, consists of a tapered, thin gage, steel body, 3.5 inches in diameter. It is cylindrically shaped and contains a shaped charge consisting of about two pounds of composition B held in place by a thin gage metal cone. When detonated, the force and heat of this explosive are focused by the metal cone, forming a small but powerful jet. The forward end of the head, called the ogive, is made of thin metal and is hollow. The ogive holds the shaped charge at the required distance from the target to obtain the maximum effect from the jet. This distance is called standoff. The jet penetrates the target and, in the case of armor, may cause some small particles to be knocked off the inside surface. If the jet hits the engine or the ammunition storage, it will probably start a fire and cause an explosion.

        (b) The HEAT rocket is used primarily against armor. It can be used against secondary targets such as gun emplacements, pillboxes, and personnel with excellent results. It is capable of penetrating heavy armor at angles of

impact greater than 30°. In an antipersonnel role, it
has a fragmentation area 10 yards wide and 20 yards
deep.
(2) *Practice rocket.*
  (a) The practice rocket (M29 series) is the same size and
  weight as the HEAT rocket and has the same flight characteristics. The head of the practice rocket is completely inert and consists of a hollow cast iron body of the
  same dimensions as the HEAT head. The weight of the
  cast iron body compensates for the absence of the filler.
  (b) The practice rocket can be fired at buttoned-up, modified
  target tanks without danger to the tank crews.

26. Technical Characteristics

(figs. 18 and 19)
a. *Motor Assembly.*
  (1) *Description.* All 3.5-inch rockets employ the same motor
  assembly. It consists of a metal tube which houses the
  propellant and igniter. The fin assembly is securely attached
  to this tube. The front end of the tube is assembled to the
  base of the fuze. The rear end forms a nozzle. The cylindrical motor cavity is divided into four sections by two
  spacer plates which support the grains of propellant powder.
  (2) *Propellant.* The propelling charge consists of 12 grains of
  M7 propellant. Each grain is 5 inches long and approximately ⅜ inch in diameter. Three grains are placed in
  each of the four sections formed by the spacer plates. Each
  lot of propellant is adjusted at the time of manufacture to
  give standard velocity. The propellant is ignited by igniter
  M20.

  Caution: **Do not fire rockets whose temperatures are
  beyond the limits marked on each rocket. The rate of
  burning increases with the initial temperature. Firing at
  temperatures below the minimum (−20° F.) gives erratic
  ranges and excessive backblast of powder particles. Firing
  at temperatures above the maximum (+120° F.) causes
  dangerous pressure within the motor.**
  (3) *Igniter and leads.* The igniter, M20, consists of a short,
  cylindrical plastic case containing a small black powder
  charge and an electrical squib. It is assembled in the forward end of the motor on top of the propellant spacer
  plates. The leads of the electrical squib, running parallel
  to the grains of propellant, pass from the igniter through
  the nozzle into the expansion cone. A green lead (ground)
  wire is connected to the aluminum support ring of the contact ring assembly. A red lead (positive) wire is attached

to a pin which is insulated from the support ring, but is in contact with the copper contact band. These connections are positioned 180° apart.

(4) *Fin assembly.* The fin assembly consists of six aluminum alloy fins and a contact ring assembly. The contact ring assembly which encircles the fins, consists of three rings. An aluminum support ring, which is innermost, is separated from the copper contact ring by a plastic insulating ring. The fins are spot welded to the expansion cone, and the expansion cone is press-fitted to the rear of the motor tube.

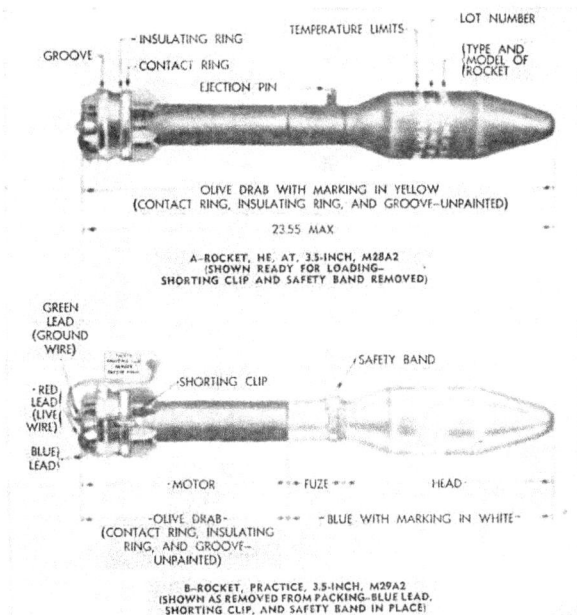

Figure 18. 3.5-inch rockets, M28A2 and M29A2.

b. *Fuze, Rocket, BD, M404 (T160E6).*

(1) *Description.* This base detonating fuze, used with the HEAT rocket, is of the simple inertia type and functions with a nondelay action upon impact. The fuze mechanism consists of a plunger, an actuating sleeve, a firing pin, a setback sleeve, a creep spring, a stop pin, and a lockpin. The explosive train includes a detonator and booster. The fuze nomenclature, the loader's lot number, and the month and year of loading are stamped into the metal. An ejection pin (boreriding safety pin), with an ejection pin spring, passes through the fuze body and prevents movement of the internal parts and accidental functioning during shipping,

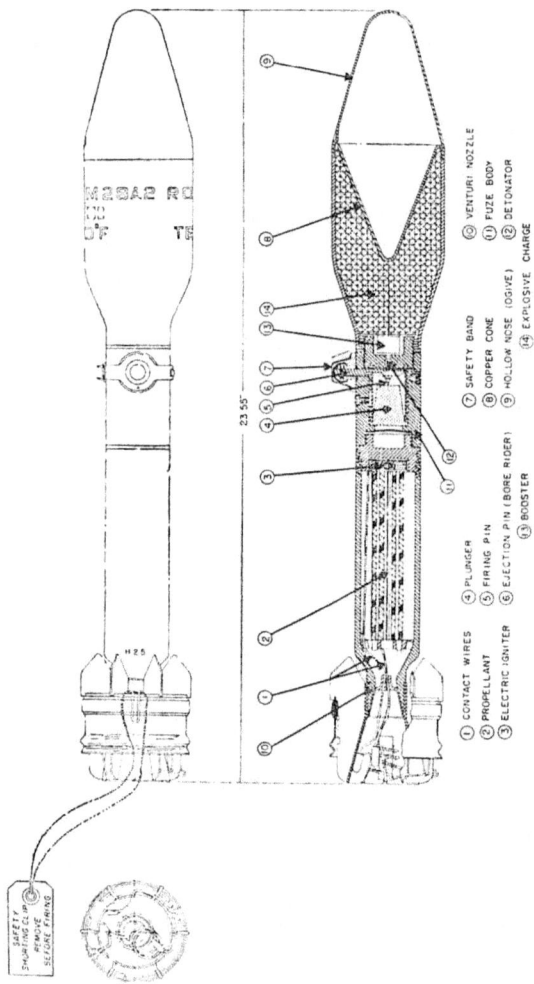

Figure 19. Cross section of rocket, M28A2 HEAT.

① CONTACT WIRES   ④ PLUNGER   ⑦ SAFETY BAND   ⑩ VENTURI NOZZLE

② PROPELLANT   ⑤ FIRING PIN   ⑧ COPPER CONE   ⑪ FUZE BODY

③ ELECTRIC IGNITER   ⑥ EJECTION PIN (BORE RIDER)   ⑨ HOLLOW NOSE (OGIVE)   ⑫ DETONATOR

   ⑬ BOOSTER   ⑭ EXPLOSIVE CHARGE

handling, and firing. An additional safety feature is provided by the safety band which prevents the ejection pin from moving during shipping and handling. The safety band is not removed from the fuze until the rocket is ready for loading.

(2) *Functioning.* When the safety band is removed, the ejection pin moves outward approximately ⅜ of an inch but still prevents all parts of the fuze mechanism from moving. When the rocket is in the firing chamber, the ejection pin is partially depressed by the chamber, thereby freeing the setback sleeve so it can move to the rear when the rocket is fired. The fuze is still safe, since the ejection pin prevents movement of the actuating sleeve and firing pin. If it becomes necessary to remove the rocket from the launcher, the ejection pin will move outward and reengage the setback sleeve. This returns the fuze to its original safe condition. When the rocket is fired, the force of inertia causes the setback sleeve to move rearward. It is held in its rearward position by the lockpin. When the rocket leaves the muzzle of the launcher, the ejection pin is thrown clear of the fuze by the ejection pin spring. The fuze is then fully armed. During flight, the firing pin lever and firing pin spring prevent the firing pin from striking the detonator. The creep spring retards the forward movement of the plunger and actuating sleeve. The action of the creep spring prevents the fuze from firing should the rocket strike light objects such as thin brush or undergrowth. Upon impact with a more resistant object, the plunger and actuating sleeve move forward until the sleeve hits the firing pin lever. This causes the firing pin to strike the detonator and explode the rocket.

c. *Fuze, Rocket, Dummy, M405 (T2008E2).* This fuze is used in the practice rocket and is inert. It incorporates an ejection pin assembly simulating that used in the BD fuze (T160E6). The body of the fuze and the safety band are painted blue. The fuze nomenclature, the loader's lot number, and the month and year of loading are stamped into the metal.

27. Identification of Rockets

A rocket is identified as to type by the color and marking of the warhead, its standard nomenclature, and its ammunition lot number.

a. *Painting.* Rocket ammunition is painted to prevent rust and to provide a ready means of identification as to type. Lusterless paint

25

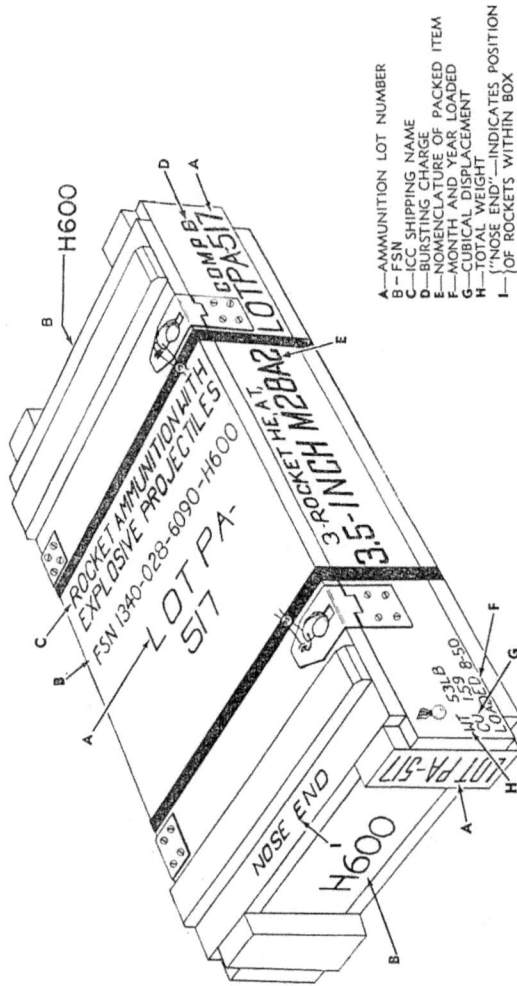

H600

COMP B
LOTPA517

A

D

ROCKET AMMUNITION WITH
EXPLOSIVE PROJECTILES

FSN 1340-028-6090-H600

ROCKET H.E.A.T.
3.5-INCH M28A2

LOT PA-
517

E

C

B

A

F

53LB
WT 159 8-50
CU.FT.
LOADED

G

H

LOT PA-517

NOSE END

H600

A

B

B

A — AMMUNITION LOT NUMBER
B — FSN
C — ICC SHIPPING NAME
D — BURSTING CHARGE
E — NOMENCLATURE OF PACKED ITEM
F — MONTH AND YEAR LOADED
G — CUBICAL DISPLACEMENT
H — TOTAL WEIGHT
I — "NOSE END"—INDICATES POSITION
      OF ROCKETS WITHIN BOX

Figure 20. Packing box for 3.5-inch HEAT rocket.

26

is used to meet requirements for camouflage. The color scheme is as follows:

ROCKET HEAD:
    HEAT_____ Olive drab, marking in yellow
    Practice_____ Blue, marking in white
FUZE:
    HEAT_____ Olive drab
    Practice_____ Blue

b. Ammunition Lot Number. When ammunition is manufactured, an ammunition lot number is assigned in accordance with certain specifications. This becomes an essential part of the marking. The lot number is stamped or marked on every loaded, complete rocket and on all packing containers. It is used for records and for reports on condition, functioning, and accidents, should they occur. In any one lot of ammunition, the components used in the assembly are manufactured under as nearly the same conditions as possible. To obtain greater accuracy when firing, one ammunition lot should be used when ever practicable.

c. Standard Markings. The standard markings are stenciled on the rocket in the appropriate color. They include type, size, model of the rocket, and the ammunition lot number. The lot number includes the loader's initial or symbol, the loader's lot number, and the date (month and year) of loading. Safe temperature limits are marked on the rocket head.

28. How Rockets Are Carried

Seven 3.5-inch rockets are carried by each rocket launcher team. The gunner carries one rocket and the loader (assistant gunner) carries six.

29. Packing

Rockets are packed in individual hermetically sealed metal containers, in quantities of three containers (three rockets) per wooden box (fig. 20). The box is marked as indicated in figure 20. In the case of practice rockets, stripes with the correct color are painted on the box. The exterior of the metal container is painted and marked with the same color scheme as used for the rocket. The marking on the container includes Federal stock number, nomenclature of the packed item, and ammunition lot number. Fiber containers are sometimes used instead of metal. Tape on these containers identifies the contents.

30. New Series Developed

A new series of rockets has been developed for the 3.5-inch rocket launcher. This series consists of the M35A1 HEAT rocket and the M36 TP rocket. For ballistic characteristics and technical information pertaining to this rocket, refer to TM 9–1950.

CHAPTER 5

PREPARATORY MARKSMANSHIP TRAINING

31. Purpose and Conduct of Training

a. Purposes.

(1) Preparatory exercises train gunners in the essentials of marksmanship and develop fixed and correct habits of marksmanship before range practice begins. The training is divided into five steps: aiming, trigger squeeze, positions, loading and immediate action, and tracking moving targets. Each step depends on material covered in the steps preceding it, and each step is essential to the final result, expert marksmanship.

(2) A prerequisite for the potential gunner is qualification in rifle marksmanship. The soldier who is not proficient with the rifle will probably not make a good gunner. The errors he makes on the rifle range will be magnified with the rocket launcher.

b. Conduct of Training. For information pertaining to the training program on the 3.5-inch rocket launcher, refer to Army Subject Schedule 21–30.

32. Aiming

a. Reticle Pattern. The tree-type sight reticle in the reflecting sight (fig. 21), consists of a broken vertical center line and broken horizontal lines. Each space and each mark on the vertical line indicate a 50-yard graduation. This gives a range selection from 0 to 450 yards without changing the setting on the elevation plate. Each space and each mark in a horizontal line indicate one lead or five MPH. Since there are three spaces and three marks on each side of the vertical line, a total lead of 30 miles-per-hour can be made in either direction.

b. Use of Tree-Type Reticle. To fire at any target between 0 and 450 yards range, set the indicator arm at the mark on the elevation plate which reads 0 to 450. Then select the horizontal line with the desired range and use it for aiming. If the target has a range greater than 450 yards, set the indicator arm on the desired range and use the zero horizontal line for aiming.

```
O   ▬   ▬   ▬   │   ▬   ▬   ▬   O

100 ▬   ▬   ▬   │   ▬   ▬   ▬   100

200 ▬   ▬   ▬   │   ▬   ▬   ▬   200

300 ▬   ▬   ▬   │   ▬   ▬   ▬   300

400 ▬   ▬   ▬   │   ▬   ▬   ▬   400
                │
```

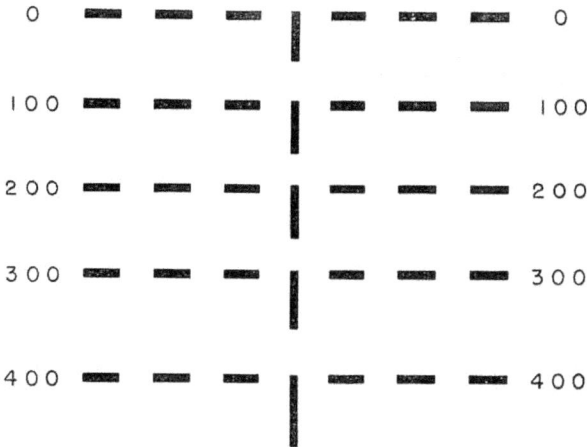

Figure 21.   Tree-type reticle pattern.

c. Stationary Targets. To fire at a stationary target, estimate the range and set the indicator arm in the correct position. Find the point at which the vertical line and correct horizontal line for the desired range cross. Place this aiming point on the center of mass of the target (fig. 22) and fire. At shorter ranges, the aim may be shifted to the most vulnerable spot on the target, such as the embrasure of a pillbox or a lightly armored part of a tank. The importance of the correct sight picture cannot be overemphasized. Use the horizontal and vertical lines to avoid canting the launcher when aiming.

d. Moving Targets. To hit a moving target, estimate the range, speed of the target, and the angle the target is approaching or leaving your position. Place the indicator arm in the correct position on the elevation plate. Then position the target in the reticle to allow for the change in deflection and range between the position of the target when the rocket is fired and the position of the target at the point of impact. See also paragraphs 57 and 58 for information on range and speed estimation.

(1) *Targets moving directly across the front.* Estimate the range and set the indicator arm at the correct position on the range scale plate. Determine the number of leads from the estimated speed of the target. Position the target in the sight so the proper lead graduation on the selected horizontal range line is on the center of mass of the target. The vertical centerline should be in front of the center of the target (figs. 23 and 24).

(2) *Targets moving directly toward or away from the gunner.* The gunner estimates the range and sets the indicator arm at the correct position on the range scale plate, disregarding

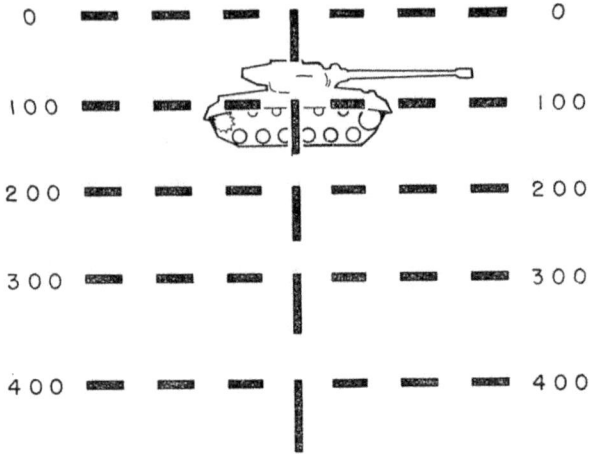

Figure 22. Stationary target 100 yards away.

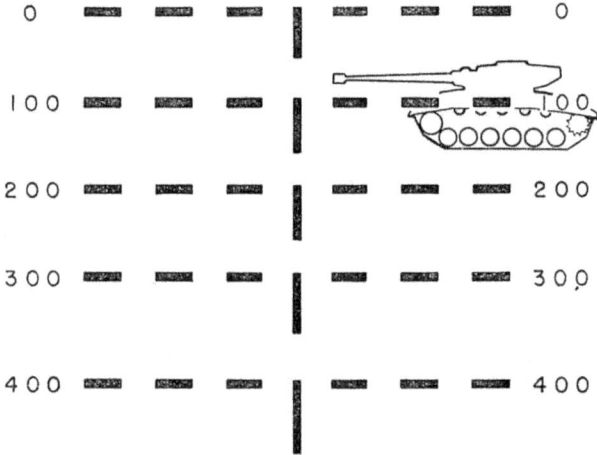

Figure 23. Target, 100 yards away, moving 30 mph right to left.

the speed of the tank. He positions the target on the reticle so the vertical centerline passes through the center of mass of the target. If the target is moving toward him, he places the selected horizontal range line on the bottom of the target; moving away from him, he places the selected horizontal range line on the top of the target (figs. 25 and 26). This compensates for the slow muzzle velocity of the rocket and takes into account the gradual range change.

(3) *Targets moving at an angle across the front.*

(a) When engaging targets moving at an angle across his front, the gunner positions the target on the reticle to account

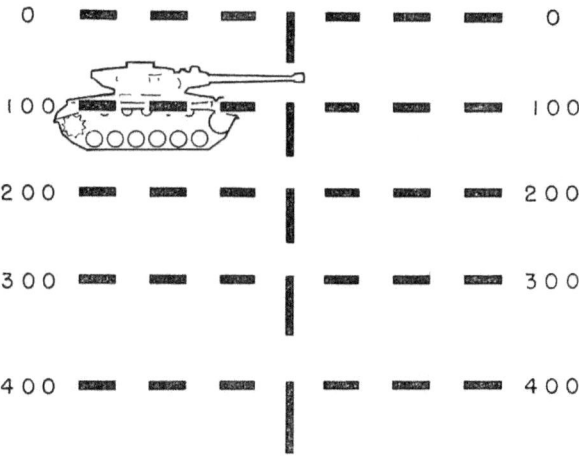

Figure 24. Target, 100 yards away, moving 20 mph left to right.

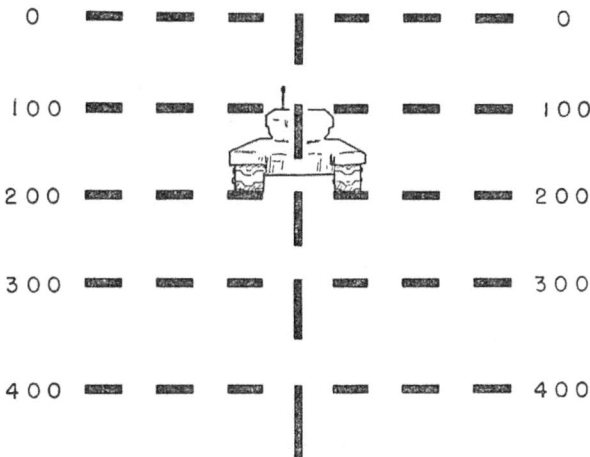

Figure 25. Target, 100 yards away, moving directly away from the gunner.

for both the movement toward or away from his position and also for the lateral movement of the target.

(b) The gunner estimates the range, speed, and angle of approach or departure of the target. He sets the indicator arm to the correct position on the elevation plate.

1. If the angle is from 0° to 45°, the gunner disregards the speed and places the selected range line either on top or at bottom of mass to compensate for movement toward or away from his position, and places the center vertical line on the leading edge to allow for the lateral movement of his target (fig. 27).

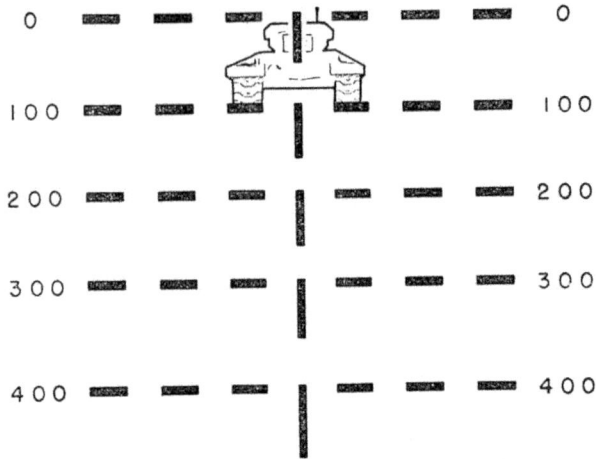

Figure 26.   Target, 100 yards away, moving directly toward the gunner.

    2. If the angle is from 45° to 90°, the gunner again places
the selected horizontal range line either on top or at
the bottom of mass to compensate for range change,
and allows for lateral movement by dividing the speed
in half and applying the appropriate lead (figs. 28
and 29).

    e. Aiming Exercise. This exercise provides practice in correct
sight alinement. The instructor places a launcher rest 1,000 inches
from, and on line with, a blank sheet of paper mounted on a box.
Then he places a launcher on the rest and points it toward the sheet of
paper. A marker with a tank silhouette target is stationed at the
box. The target is approximately 5 inches long and 2 inches high.
Without touching the launcher on the rest, the instructor takes his
position and looks through the sight. The marker holds the target
(tank silhouette) against the paper and moves it as directed by the
instructor. When the sight picture is correct, the instructor commands HOLD. Then he moves away from the launcher without dis-
turbing the sight and directs the gunner to look through the sight to
see the correct sight picture. After the correct sight picture is ob-
served, the marker moves the target out of alinement. The gunner
takes his position at the launcher and directs the movements of the
target until the correct sight picture is obtained. He repeats this
operation three times. After each attempt, the marker makes a dot
on the paper through a hole in the center of the target. The three
resulting dots should form a shot group not over 1 inch long on the
longest axis. This exercise is repeated with sight settings of 200 yards
with no lead; 100 yards, one lead; and 150 yards, two leads. Each
exercise is repeated until satisfactory shot groups are obtained.

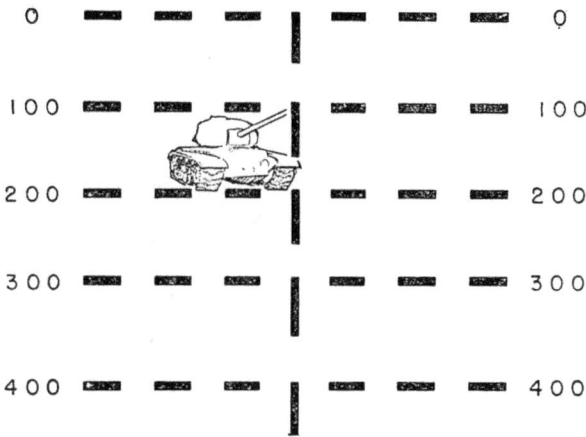

Figure 27.   Target, 200 yards away, approaching from the left front at an angle of 30°

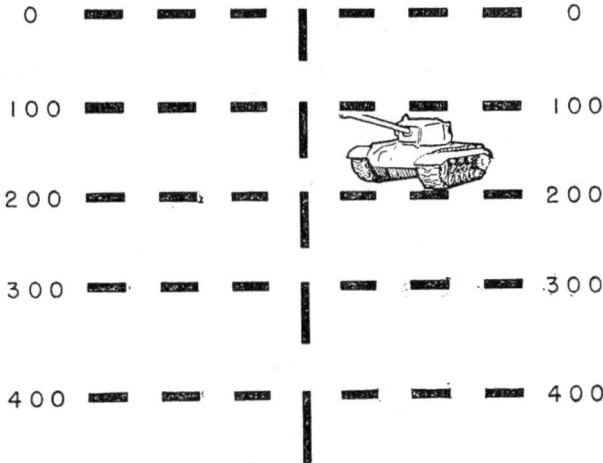

Figure 28.   Target, 200 yards away, approaching from the right front at an angle of 45° and a speed of 20 mph.

This gives trainees practice in precise aiming under different conditions of range and speed.

33. Trigger Squeeze

The gunner places the first and second fingers of his right hand on the trigger and alines the sights.   When the correct sight picture has been obtained, he squeezes the trigger with a smooth, steady, rearward pressure.   He holds his breath while aiming and squeezing the trigger.   After squeezing the trigger, the gunner maintains the correct sight picture until the rocket leaves the launcher.   This is a much longer period than is required for the rifle.

33

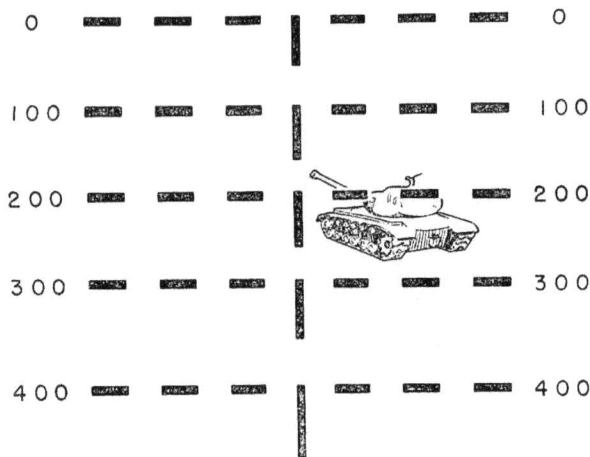

Figure 29. Target, 200 yards away, moving away from the gunner from the right front toward the left at an angle of 60° and at 20 mph.

Caution: Dropping the muzzle of the launcher immediately after firing may cause the rocket to strike the ground directly in front of the gunner. Serious injuries to personnel could result.

34. Position—General Rules

a. The rocket launcher is fired from the prone, kneeling, modified kneeling, and standing positions in a manner similar to those used to fire other shoulder weapons. The exact position is limited largely by the conformation of the body. The gunner makes sure his position is comfortable, relaxed, and steady.

b. The launcher is fired at moving targets at ranges up to 200 yards (185 M); therefore, the gunner assumes the position which gives the greatest flexibility. When firing at a moving target, he keeps his body flexible. His arms and the upper part of his body are free, permitting him to rotate at the waist. He does not rest his elbows on his knees. This provides smooth, steady, easy motion in the direction of the movement of the target, while obtaining the desired lead. For this reason, the prone position, although it offers a low silhouette, is not suitable for firing at a moving target.

c. Stationary targets may be engaged at ranges up to 300 yards (275 M) in the prone position (fig. 30). For ranges beyond 300 yards (275 M), the prone position is not recommended because of the extreme elevation of the launcher. This causes the backblast to strike the ground dangerously near the crew.

d. When assuming a position, there is a point at which the launcher feels comfortable and rests naturally without strain on a given point. In a correct position, this direction is toward the target; therefore, the gunner shifts his body until the launcher points naturally at the target area, remembering—

    (1) Not to cant the launcher.

    (2) To practice and fire from a right-hand position.

    (3) To keep the muzzle up and toward the target at all times.

35. Prone Position
    (fig. 30)

   a. Gunner. To take the prone position and to avoid injury from the backblast of the weapon, the gunner—

    (1) Lies at an angle of not less than 45° to the line of aim, as shown in figure 30.

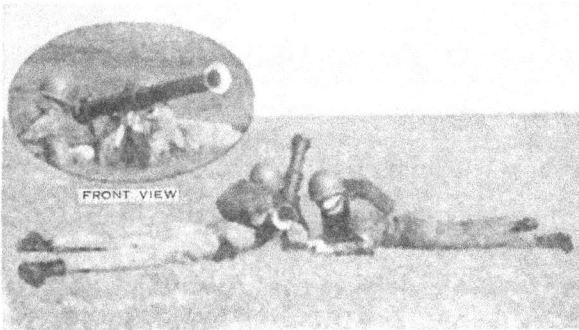

Figure 30. Prone position.

    (2) Keeps his spine straight. In this position his right leg is directly on a line running through his right hip and right shoulder.

    (3) Moves his left leg to the left as far as possible with comfort.

    (4) Holding his elbows well under the launcher, cradles the trigger guard with his left hand.

    (5) Places the stock firmly against his shoulder.

    (6) Holds his head as steady and comfortable as possible, lining up his right eye with the sights.

    (7) If forced to track a moving target from this position, maintains the 45° angle so he does not accidentally place his feet in the path of the backblast. The prone position is not suitable for tracking.

   b. Loader. The loader takes his position flat on the ground opposite the gunner's right shoulder, perpendicular to the weapon, and facing slightly to the rear. He—

    (1) Rests himself on both elbows.

    (2) Takes his position close enough to the gunner to be able to communicate with him, and at the same time load the launcher.

35

(3) Moves about to conform with the gunner's movements, to avoid backblast, and to load the launcher.

36. Kneeling and Modified Kneeling Positions

a. Gunner. There are two kneeling positions: one for stationary targets and one for moving targets.

(1) To assume the modified kneeling position for moving targets (fig. 31), the gunner kneels on his right knee with the upper part of his right leg vertical. His buttocks do not rest on his right heel. He points his leg toward the target with his left foot at a right angle to and opposite his right knee. His left leg forms a right angle to the ground. Holding his body erect, he places his left elbow under the piece. He cradles the trigger guard in his left hand, grasping the hand-grip with his right hand. He holds his right elbow high, forming a pocket for the stock.

(2) The kneeling position for the stationary target (fig. 32) is similar to the kneeling position for firing the rifle. The gunner kneels on his right knee at a 90° angle with the line of aim, sits so the center of his right buttock rests directly on his right heel, and shifts his weight forward so his heel inclines in the direction of the target. When viewed from the front, his left leg is vertical; however, it need not appear vertical when viewed from the side. The gunner may prefer to draw his left foot back, relaxing his body weight forward so a solid contact is made between his calf and thigh. He points his left toe in the direction which gives him the most comfort, rests his left upper arm on his left knee, and places his left and right hands as prescribed for the modified

Figure 31. Modified kneeling position for firing at a moving target.

Figure 32. Kneeling position for firing at a stationary target.

kneeling and standing positions. He raises his right elbow to the height of, or slightly above, his shoulder, forming a pocket for the stock.

b. Loader. The loader takes a kneeling position opposite the gunner's right shoulder and faces to the rear of the launcher (figs. 31 and 32). He places one or both knees on the ground. If he places one knee on the ground, it must be the knee closest to the breech. He places himself close enough to the gunner to communicate with him, and at the same time load the weapon. As in the other positions, the loader moves about while tracking to conform with the movements of the gunner, to avoid burns, and to load the weapon.

37. Standing Position

a. Gunner. The standing position for using the launcher is similar to that used for firing the rifle (fig. 33). The gunner stands half-faced to the right with his feet a comfortable distance apart and his body erect and well balanced. He places his left elbow comfortably under the launcher and raises his right hand, cradling the trigger guard in his left hand. To traverse in this position, he moves his body from the ankles up.

b. Loader. The loader takes a standing position opposite the gunner's right shoulder (fig. 33). He faces the rear of the launcher, close enough to the gunner to load the weapon. While tracking, he moves about to conform with the gunner's movements, to prevent burns, and to load the weapon.

38. Loading Launcher

a. At all times during loading, aiming, and firing, the loader is to the side and slightly forward of the breech end of the launcher.

Figure 33. Standing position.

The gunner is in command of the rocket team. Training in loading and immediate action is conducted concurrently.

    b. The gunner—

        (1) Checks the bore of the launcher to make sure it is clean.

        (2) Assumes the prone, kneeling, modified kneeling, or standing position.

        (3) Places the desired range setting on the sight assembly.

        (4) Places the launcher on his right shoulder pointed at the target area.

        (5) Sets the SAFETY in the SAFE position.

        (6) Removes his right hand from the trigger and supports the launcher with his left hand under the trigger guard.

        (7) Taps the loader with his right hand and at the same time commands LOAD. During early training periods and until he is throughly familiar with the launcher, the gunner places his right hand on his helmet. This is a safety measure to prevent him from squeezing the trigger.

c. The loader, when loading the launcher—

(1) Takes a position similar to the gunner's, but on the opposite side of the launcher. He places himself so he faces, and is within easy arm reach of, the breech. During loading, aiming, and firing, he does not stand behind the launcher.

(2) Repeats the command LOAD.

(3) Without changing his position, picks the rocket up with his left hand, palm up. He points the head of the rocket toward the target.

(4) Checks the rocket for a loose nozzle closure by gently pulling the red and green ignition wire leads which pass through the closure. Any movement of the closure indicates it has not been sealed properly. A loose closure may result in the rocket falling short or "chuffing" (intermittent burning with a puffing noise) when fired. A chuffing rocket may fall to the ground a short distance from the launcher, smolder, and then resume burning and be propelled in an unpredictable direction. Rockets with loose nozzle closures must not be fired. Special attention must be given to examining the nozzle closure during wet or freezing weather.

Warning: **Any rocket in which the head moves with respect to the fuze, or has a discernable gap between the head and fuze, is not to be fired. Any rocket evidencing either of the above conditions will be returned to segregated ammunition storage in a properly marked container indicating the condition (see par. 12 TM 9–1950).**

(5) Removes the shorting clip from the copper band (fig. 34).

(6) Removes the safety band from the rocket (fig. 35).

(7) Grasps the rocket, with ejection pin pointed down or to the side. In this position, it will not strike the position stops on the launcher when the rocket is loaded.

(8) Rotates the control handle forward to the LOAD position (fig. 36).

(9) Holds the control handle in the LOAD position and inserts the rocket in the launcher tube until it is stopped in position by the right- and left-hand stops (fig. 37).

Warning: **Never ram the rocket into the launcher. Precise, unhurried loading prevents the stops from overriding the fin assembly, or allowing the rocket to move too far forward into the barrel.**

(10) Rotates the control handle rearward to the FIRE position (fig. 38). The launcher is now ready to be fired.

(11) Glances to the rear to see that the backblast area is clear.

(12) Taps the gunner and calls UP.

Figure 35. *Removes safety band.*

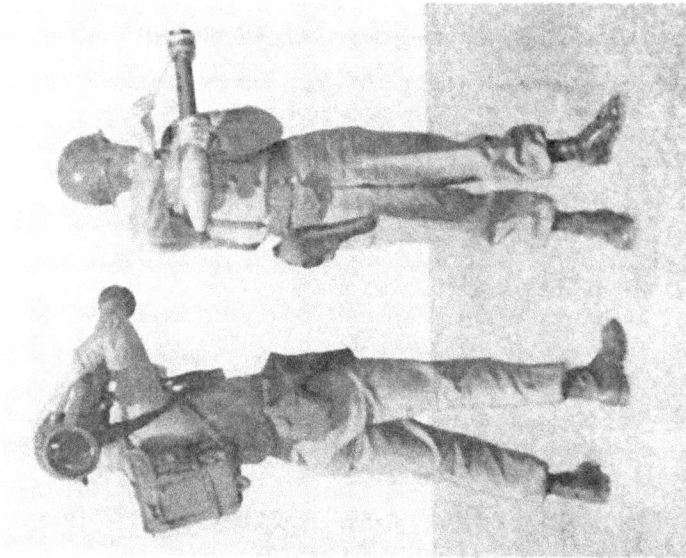

Figure 34. *Remove shorting clip.*

Figure 36. Control handle in LOAD position.

Figure 37. Control handle in LOAD position with rocket inserted.

d. The gunner takes his right hand off his helmet and places it back on the handgrip. He checks his range setting and moves the safety to the FIRE position. He alines his sights on the target and squeezes the trigger. The gunner maintains his sight picture until the rocket has cleared the tube.

Warning: **If a slight noise is heard and a small puff of smoke is emitted from the rear of the launcher during an attempt to fire, this indicates the igniter has functioned but has failed to ignite the propellant. If this is the case, it is mandatory to keep the launcher** trained on the target and to observe all the precautions for firing for a 2-minute period. At the end of this period, remove the rocket and return it to segregated ammunition storage in a properly marked container indicating the condition.

39. Malfunctions and Immediate Action

a. Malfunctions. Malfunction of the launcher is classified as a failure to function satisfactorily. Some of the more common malfunctions and the corrective measures used to reduce them in the field are as follows:

(1) *Failure to load.*

Cause	Correction
Bent tube_____	Replace the launcher.
Oversize or bent rocket	Discard and get a new rocket.

(2) *Failure to fire.*

Cause	Correction
No contact between the firing contact and the copper contact band detent and the unpainted groove of the rocket fin assembly.	Move control handle to LOAD position and reposition the rocket.
Fin assembly overrides the right- and left-hand stops.	Make sure trigger mechanism is set on SAFE. Move control handle to FIRE position. Raise the detent latch with right hand and grasp the fin assembly with left hand. Slide the rocket to the rear. Release detent latch and move the control handle to the LOAD position. Reposition the rocket in the launcher.

b. Immediate Action. Immediate action is the prompt action taken by the firer to reduce a stoppage. If a misfire occurs, the following immediate action is performed:

(1) The gunner maintains the sight picture and releases the trigger. He then squeezes the trigger two additional times. If the rocket still fails to fire, he moves the safety to SAFE position, drops his right hand from the trigger group, taps the loader, and calls MISFIRE.

> *Note.* The gunner does not lower the muzzle before releasing the trigger because this may fire the rocket into the ground to his immediate front.

(2) The loader repeats MISFIRE. He waits 15 seconds, counting slowly to 15 to allow for a possible hangfire. He ro-

Figure 38.   Control handle in FIRE position.

tates the control handle to LOAD position and checks the detent to see it is correctly engaged in the unpainted aluminum groove of the rocket fin assembly.   At the same time, he rotates the rocket clockwise to insure electrical contact between the copper contact band and the firing contact which is between the detent and unpainted aluminum groove. He rotates the control handle rearward to the FIRE position, taps the gunner, and calls UP.

(3) The gunner relays and attempts to fire.   If the launcher still fails to fire, he repeats the procedure for immediate action for a misfire.   If the rocket still does not fire, he commands UNLOAD.

(4) The loader repeats UNLOAD and waits 15 seconds, counting slowly to 15.   He moves the control handle to LOAD position and withdraws the rocket from the launcher in the following manner: he rests the palm of his right hand on the breech guard, placing his fingers on the detent and lifts up on the rocket; he grasps the rocket by the fin assembly and, with his left hand, removes it; he replaces the safety band over the ejection pin and shorting clip on the copper contact band; and then he puts the rocket aside for repacking and disposal.

(a) If, while removing the rocket from the launcher, the ejection pin tends to bear against the breech guard of the launcher, the loader presses the pin back and holds it. As soon as the rocket has been removed from the launcher, he installs the safety band over the ejection pin.

(*b*) Without the ejection pin in place, the rocket is armed and must be handled *nose up* with extreme care. If the ejection pin cannot be installed, the rocket is destroyed as soon as possible in accordance with TM 9–2002. The loader never stands directly behind the launcher.

(5) If the rocket fails to fire, the loader unloads the launcher and checks the spring action of the firing contact. He checks the detent for positive ground contact. He cleans both the firing contact and detent and traces the electrical circuit of the launcher. He looks for broken wiring and insures the rivet makes contact with both the right- and left-hand contact brackets when in the firing position. He checks the spring action of the right- and left-hand stops. Stops should protrude into the bore when in LOAD position.

40. Tracking Moving Targets

a. This phase of training helps the gunner maintain the correct sight picture when tracking a moving target. A moving vehicle or a towed target is used for this training. The targets are operated to the front at different ranges, at varying speeds, and from many directions.

b. The gunner assumes his position, sets the correct range on the sight, and tracks the target with the correct sight picture. He tracks the target smoothly and continuously. In practice and when firing, he continues to watch the target while his launcher is being reloaded.

CHAPTER 6

MAINTENANCE

41. Disassembly and Assembly of Launcher

The launcher is disassembled only for inspection, maintenance, and repair, and only by organizational maintenance personnel at those times. See TM 9–2002 and TB 9–2002–1. Disassembly for ordinary care and cleaning is unnecessary.

42. Operational Inspection

a. General. An operational inspection of the launcher is made at delivery and periodically thereafter unless required sooner by extensive firing or unusual weather conditions. It consists of a thorough and systematic check followed by necessary repairs and adjustments.

b. Procedure.

(1) Examine the launcher for general condition, loose or broken parts, chipped paint, bends and dents, and obstructions in the bore.

(2) Check for broken or loose wire connections.

(3) Look for corrosion on the electric contact points.

(4) Test the functioning of the contactor latch group assembly.

(5) Examine the barrel of the launcher for rust, scale, and dents.

(6) Examine the barrel coupling screw and nut for burs. Test the functioning of the barrel coupling lock, and check for looseness in the barrel coupling.

(7) Check the sight for ease of folding and extending and for loose or broken lenses.

(8) Remove the trigger grips. Look for loose connections at the soldered points and check the magnets and coil for looseness. Test the functioning of the trigger spring, rocker arm spring, and armature spring. (This final portion of the inspection is made only by the unit maintenance personnel.)

43. Preventive Maintenance

a. The rocket launcher crews—

(1) Inspect the launcher daily as outlined in paragraph 42*b* (1) through (5). If defective, turn in for repair or replacement.

(2) Cover all bearing surfaces with a light coat of preservative lubricating oil (special), using an oiled cloth.

(3) Coat the barrel with a light coat of preservative lubricating oil (special), except when firing.

(4) Lubricate the firing mechanism weekly by injecting a few drops of preservative lubricating oil (special) through the rocker arm toggle pin hole. Lift the trigger spring and washer slightly to expose the opening on the rocker arm cover for the toggle pin.

b. The unit maintenance personnel—

(1) Inspect the launcher periodically as outlined in TM 9–2002 and TB 9–2002–1.

(2) Tighten loose parts, replace broken parts, and keep the paint in good condition.

(3) Remove traces of rust from unpainted surfaces with a crocus cloth.

(4) Notify Ordnance maintenance personnel when parts of the firing mechanism are worn or broken.

(5) Remove the grips and lubricate the electrical firing mechanism by coating the entire mechanism with a film of oil.

(6) Wipe off excess oil and assemble the grips to the firing mechanism.

44. Repairs

Unit maintenance personnel make the following repairs:

a. Solder all loose or broken soldered connections.

b. Replace broken or cracked trigger grips.

45. Lubrication Order

a. The lubrication order LO 9–2002 (fig. 39) prescribes organizational and higher echelon maintenance lubrication for this equipment.

b. One lubrication order (fig. 39) is issued with each item of equipment. It accompanies the item at all times. If equipment is received without copies, the using organization immediately requisitions a replacement. See DA Pam 310–4.

c. Instructions on lubrication orders are binding on all levels of maintenance.

d. Service intervals specified on lubrication orders are for normal operating conditions and during active service. These intervals are reduced under extreme conditions such as excessively high or low temperatures, prolonged periods of firing, sandy or dusty locations, immersion in water, or exposure to moisture. These conditions may destroy the protective qualities of the lubricant.

e. The grades of lubricants are prescribed in the lubrication order (fig. 39).

LUBRICATION ORDER **LO 9-2002**

24 September 1956 (Supersedes LO 9-2002, 13 Jan 56)

LAUNCHER, ROCKET, 3.5-IN M20, M20A1, M20B1 AND M20A1B1

References: TM 9-2002, ORD 7 SNL B-42

BORE—After firing and on 2 consecutive days thereafter clean with CR. After 3d cleaning, wipe dry, oil with PL Special above 0°F or LAW below 0°F. Wipe clean before firing. When weapon is not being fired, renew oil film weekly.

BARREL COUPLING LOCK AND CONTACTOR LATCH—After firing and weekly thereafter oil with PL Special above 0°F or LAW below 0°F.

FIRING MECHANISM—(C7313326) or (B7140325) lubricated by organizational mechanics only.

PL Special—Lubricating Oil, General Purpose
LAW—Lubricating Oil, Weapons
CR—CLEANING COMPOUND, SOLVENT:
Rifle bore cleaner

By Order of
WILBUR M. BRUCKER
Secretary of the Army

MAXWELL D. TAYLOR
General, United States Army
Chief of Staff

OFFICIAL:
JOHN A. KLEIN
Major General United States Army
The Adjutant General

INSTRUCTIONS CONTAINED HEREIN ARE MANDATORY

Figure 39. Lubrication order.

46. Special Lubrication Instructions

a. Bore. Clean and oil as prescribed in the lubrication order (fig. 39).

(1) Place rifle-bore cleaner on a wiping cloth and run the cloth through the bore several times. Then run a clean, dry wiping cloth through the bore.

(2) Repeat these operations until the bore is clean, then apply a film of rifle-bore cleaner.

(3) For two consecutive days after firing, clean the bore as described in (1) and (2) above. On the third day, dry the bore thoroughly and oil as prescribed in the lubrication order.

b. Contactor Latch Group Assembly. Clean latch assembly. Oil lightly with preservative lubricating oil (special) all rivets, the pins upon which the linkage assembly pivots, the firing contact, and the cam surface (fig. 11) of the linkage assembly.

c. Electrical Firing Mechanism (fig. 40). The organizational mechanic is responsible for lubricating the firing mechanism. He—

(1) Removes and cleans the grips monthly.

(2) Applies the prescribed lubricant to the trigger pin and to the slot in the back of the trigger above and below the trigger latch. He operates the electrical firing mechanism several times after applying lubricant to the trigger latch.

(3) Applies the prescribed lubricant to the roller on the trigger latch, to both pins in the trigger bar, and to the armature spring sleeve.

(4) Places the blade of a screwdriver against the front of the top plate of the firing mechanism so the safety lugs of the trigger stop about ⅛ inch from the top plate when the trigger is squeezed. He holds the trigger lugs against the screwdriver. This keeps the top of the armature in its rearward position and also exposes the slot in the armature spring sleeve and a little of the armature spring. He applies the prescribed oil to the opening in the sleeve, squeezing the trigger a few times to work the oil into the sleeve.

(5) Wipes off excess oil and assembles the grips to the firing mechanism.

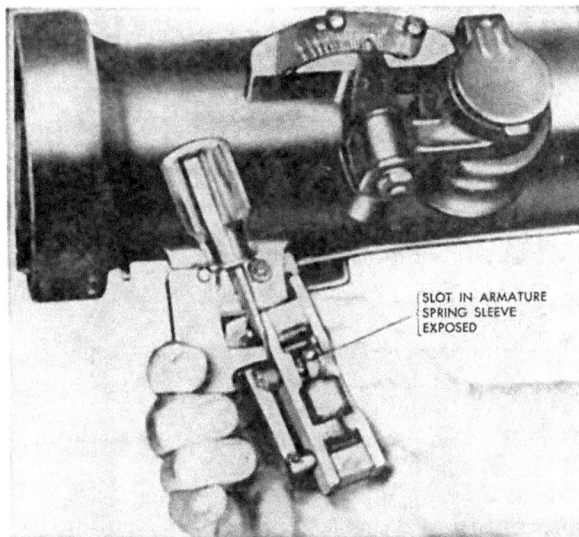

SLOT IN ARMATURE
SPRING SLEEVE
EXPOSED

Figure 40. Lubrication of armature spring in electrical firing mechanism.

47. Operating Under Unusual Conditions

a. General.

(1) The launcher is operated the same during unusual atmospheric conditions as it is during normal conditions. However, special emphasis is placed on correct cleaning and lubrication in salty or humid atmosphere and in extreme temperatures.

(2) Under all conditions, the bore and the contactor latch housing are wiped thoroughly to remove excess oil before firing.

b. Cold Weather Preparation and Operation. Prior to cold weather operations, oil the weapon with the prescribed lubricants according to the expected temperature change. When changing the grade of lubricant, it is imperative the parts be disassembled and cleaned before the new lubricant is applied.

(1) Apply the following instructions for cold weather operations and daily care:

    (*a*) Keep moving parts and the bore of the launcher free of ice or snow.

    (*b*) Inspect material thoroughly. Protection is provided by installed and securely fastened covers. If moisture or ice is found, the affected surface must be cleaned, thoroughly dried, and coated sparingly with a light oil.

(2) Prevent condensation. If possible, do not bring the material indoors. If it is brought indoors, clean dry, and lubricate the launcher immediately. Repeat after the launcher has reached room temperature.

(3) Consider the following when firing in temperatures below freezing:

    (*a*) Give special attention to protecting the eyes, face, and hands. Rockets of the M28 and M29 series will produce afterburning at a lower operating temperature. The increase in the burning time and distance causes the rocket to burn beyond the length of the launcher.

Caution: **Face and hand protection are mandatory when firing the launcher at below freezing temperatures. Without such protection, face and hand injuries are almost certain to result. For temperatures below 70° F., the field protective mask will be used for face protection; other types of face protection are not authorized. However, after use for firing, it must be inspected by competent personnel BEFORE using for CBR training. For temperatures above 70° F., the antiflash mask is to be worn.**

    (*b*) Moisture entering the fuze cavity is injurious in normal and freezing temperatures. The safety band depresses the ejection pin and seals the fuze against moisture. When the band is removed, the ejection pin moves to the locked position, leaving an opening around the pin. Moisture can enter the fuze cavity and freeze if considerable time elapses between removal of the safety band and firing. This may prevent functioning of the fuze at the target.

Caution: **Do not remove the safety band until you are ready to fire. Keep the rocket dry if possible.**

    (*c*) Prevent ice or frost from accumulating on rockets to keep them from freezing to the launcher, and to insure accuracy of flight. The 3.5-inch rocket develops a powerful thrust when fired. If the rocket freezes to the bore and is fired, the weapon might be violently wrenched from the gunner's grasp, or the gunner and weapon may be hurled forward with possible injuries and damage. The accumulation of ice on the rocket affects its weight, balance, or the flow of air through the fins. This has an adverse effect on accuracy of flight and range.

c. Tropical Climate. In tropical climates where temperatures and humidity are high or where salt air is present, inspect the launcher daily and clean it thoroughly. In humid, salty atmosphere, lubricate the bore and all unpainted metal surfaces daily as prescribed in the lubrication order (fig. 39).

d. Operation in Sandy or Dusty Conditions. In hot, dry climates where sand and dust are apt to get into the bore, wipe all moving parts and the bore clean at least daily. Lubricants containing sand or dust form abrasive mixtures. During sand or dust storms, keep the launchers covered if possible.

CHAPTER 7

RANGE PRACTICE

48. General

Range practice follows the satisfactory completion of preparatory marksmanship training. An officer in charge of the firing is present *during all firing*. He is responsible for the conduct of the firing and the enforcement of safety.

49. Safety Precautions

(AR 385-63)

The following safety precautions apply when firing the rocket launcher:

a. Before firing a rocket, clear the area to the rear of the launcher of personnel, material, and dry vegetation as indicated for zones A and B in figure 41. Clear zone A (the blast area) of all personnel, ammunition, material, and inflammables such as dry vegetation. The danger in this zone is from the blast of flame to the rear. Clear zone B of all personnel and material unless protected by adequate shelter. The principal danger in zone B is from the rearward flight of nozzle closure and/or igniter wires. An additional safety factor for training is obtained by zone C.

b. Range firing does not begin until the officer in charge knows the range is clear. *He gives the command to fire.*

c. At least one officer is present during all firing.

d. Launchers are removed from the firing line by order of the officer in charge.

e. No person moves to or leaves the firing line without permission from the officer in charge.

f. The ready line is placed to the flank of the firing line to protect the men.

g. No one passes through the area between the rear line and the firing line without permission from the officer in charge.

h. All loading and unloading are done on the firing line with the launcher on the gunner's shoulder. *The muzzle is pointed down range; not toward the ground.*

i. Misfires are unloaded only on command of the officer in charge.

j. Ammunition is issued only on the ready line.

k. Ammunition is placed out of range of backblast.

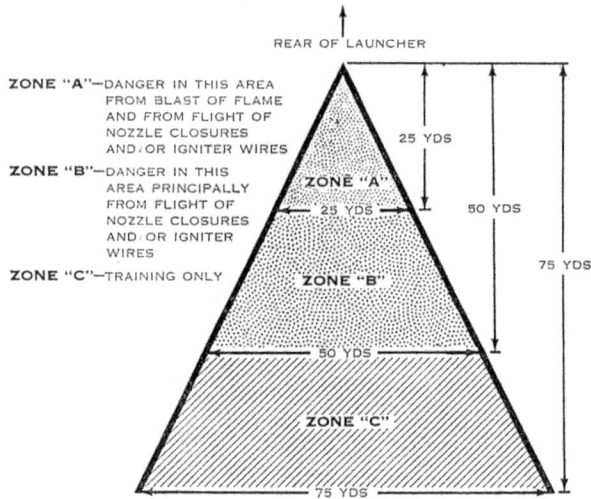

REAR OF LAUNCHER

ZONE "A"—DANGER IN THIS AREA
FROM BLAST OF FLAME
AND FROM FLIGHT OF
NOZZLE CLOSURES
AND/OR IGNITER WIRES

ZONE "B"—DANGER IN THIS
AREA PRINCIPALLY
FROM FLIGHT OF
NOZZLE CLOSURES
AND/OR IGNITER
WIRES

ZONE "C"—TRAINING ONLY

ZONE "A"

ZONE "B"

ZONE "C"

25 YDS

25 YDS

50 YDS

75 YDS

50 YDS

75 YDS

Figure 41. Danger zone to the rear of the 3.5-inch rocket launcher.

l. Ammunition is covered to protect it against the direct rays of the sun.

m. While on the ready line, each team inspects its equipment before firing.

n. Ammunition is not fired at temperatures outside the safe temperature range for that particular type of ammunition.

o. When firing all rockets from shoulder launchers, operating personnel will wear the antiflash mask at temperatures of 70° F. and above. For temperatures below 70° F., the field protective mask is worn (see par. 18*h*, TM 9–1950). These are needed to protect their eyes and faces against small particles of unburned propellant which may be blown from the rear of the rocket as it leaves the muzzle of the launcher.

p. To account for all duds when firing HEAT rockets, a noncommissioned officer is designated to count the number of rockets fired and the number of explosions. Range firing and the destruction of duds are conducted according to the provisions as outlined in AR 385–63 and TM 9–1900.

50. Boresighting

a. To score hits, the rocket launcher must be carefully boresighted before firing. Boresighting parallels the axis of the sight with the axis of the bore when a range setting of zero is on the sight. It is accomplished by the distant target or infinity method. (For the method to be used by Ordnance maintenance personnel, see TM 9–2002.)

b. The following equipment is needed to boresight the launcher:

    (1) Screwdriver.

    (2) Adjustable wrench.

    (3) Wooden disk with ⅛-inch hole in the center for use in the breech of the launcher (obtained from the ammunition container).

    (4) Two lengths of string for use as crosshairs across the muzzle of the launcher or a wooden disk with crosshairs to be placed in the muzzle of the launcher.

    (5) A solid rest (fig. 42).

c. The target on which the launcher is to be boresighted should be about 1,500 yards (1,370 meters) from the weapon.

d. The following method is used to boresight the launcher:

    (1) The muzzle deflectors of the front barrels must have score marks on the rim. These marks are spaced 90° apart. These notches are used to position crosshairs for boresighting. When these notches are missing, they must be added by Ordnance maintenance personnel.

    (2) Install two lengths of string as crosshairs across the muzzle, locating them in the score marks which appear on the rim of the muzzle. Remove the wooden disk from the ammunition container and insert it in the rear of the launcher as a boresight plug. Sight through the plug and aline the launcher so the distant target appears exactly at the intersection of the crosshairs. The launcher must be set firmly in this position.

    (3) If the muzzle deflector of the launcher does not have score marks to locate the crosshairs, position the eye at least 3 feet behind the launcher so the outline of the front opening is centered in the outline of the rear opening. At the same time, center the distant target in the outline of the front opening of the launcher. Set the launcher firmly in this position.

    (4) Set the indicator at 0 on the elevation plate. Sight through the reflecting sight assembly. If the image on the target is not at the intersection of the vertical center line and the 0 horizontal line, the reflecting sight assembly is out of adjustment. Models of more recent manufacture have a stop nut and screw located on the sight hinge (fig. 17). When the sight is in the extended position, it is prevented from pivoting beyond the extended position by the stop screw resting against the indicator arm yoke. Before adjusting for deflection, turn the stop screw counterclockwise to create a gap between it and the indicator arm yoke. After adjustment has been made for deflection, close any existing gap and tighten the stop nut. To move the sight to the

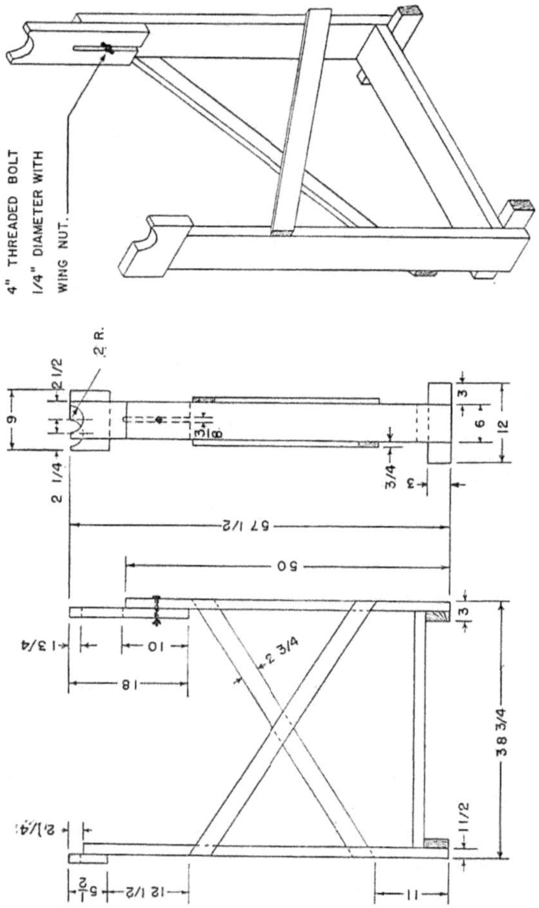

4" THREADED BOLT
1/4" DIAMETER WITH
WING NUT.

Figure 42. Launcher rest.

right or left, loosen the hinge stud nut approximately two turns with an adjustable wrench. Turn the hinge stud (fig. 17) to the desired position with a screwdriver. Hold the hinge stud in position with a screwdriver and tighten the hinge stud nut. To move the sight up or down, loosen the elevation plate screws (fig. 17), and move the elevation plate to the desired position. Tighten the elevation plate screws.

(5) The launcher is correctly boresighted when the target is centered in the bore (intersection of crosshairs) and the intersection of the vertical centerline and the 0 horizontal line in the sight is centered on the target (fig. 43).

(6) To see that the target is correctly alined in both sight and bore, it is necessary to check all three locked positions of the front barrel in the rear barrel. Do this after the final adjustment has been made. If it is not correctly alined, replace the front barrel.

51. Stationary Targets

Stationary targets 6 feet square may be constructed of logs, armor plate, or light wooden frames covered with target cloth or paper (fig. 44).

52. Moving Target

A modified medium tank may be used as a moving target. When a tank is not available, use a wooden frame target, 6 feet by 18 feet, and cover it with target cloth or heavy paper (fig. 45). Mount the target on a sled and tow it, using a vehicle (fig. 46). The construction of the sled and target is illustrated in figure 45. Any other type of moving target, which has the same dimensions and is capable of attaining the speeds called for in table II, may be used.

53. Instruction Practice

After satisfactorily completing preliminary training, fire the course outline in tables I and II with the practice rocket (M29 series). After the launcher has been boresighted, two rockets are allowed for zeroing before the gunner fires the qualification course.

54. Qualification Firing

a. Qualification Course. On completion of the instruction practice, fire tables I and II for qualification with the practice rocket (M29 series).

b. Qualification Scores.

Rating	Qualification scores
Expert_____	6-7
Gunner, 1st Class_____	4-5
Gunner, 2d Class_____	3
Unqualified_____	Less than 3

Table I. Stationary Target

Position	Range	No. of rockets	Score
Kneeling_____	100 yds_____	1	1
Prone_____	200 yds_____	1	1

Table II. Moving Targets

Target run	Position	Range	Distance between limiting stakes	Speed of target	Direction of target	No. of rockets	Score
1	Kneeling_	100 yards_____	50 yards_____	15 mph_	L to R_	1	1
2	Standing_	100–150 yds (diagonal).	71 yards_____	10 mph_	R to L_	1	1
3	Kneeling_	150–100 yds (diagonal).	71 yards_____	10 mph_	L to R_	1	1
4	Standing_	150 yards_____	100 yards_	10 mph_	R to L_	1	1
5	Kneeling_	200 yards_____	50 yards_____	10 mph_	L to R_	1	1

c. Officer in Charge of Firing. The officer in charge of firing is responsible for the conduct of all record practice. The officers needed to supervise the record firing are detailed as assistants to the officer in charge of firing.

d. Range Procedure.

(1) An officer or designated noncommissioned officer scores each man firing for record.

(2) For firing target runs 2 and 3, the diagonal target run, the firing points are spaced at 5-yard intervals. Every third firing point is used on one target run.

(3) Each firer has a loader to assist him on the firing line. During firing, the loader completes the loading operation and taps the gunner to indicate loading is complete. This is the extent of his assistance to the gunner.

(4) The ranges and the course of the moving targets are shown in tables I and II. A distance of 25 yards between stakes A–B and E–F (fig. 46) is allowed to permit the tank to reach

56

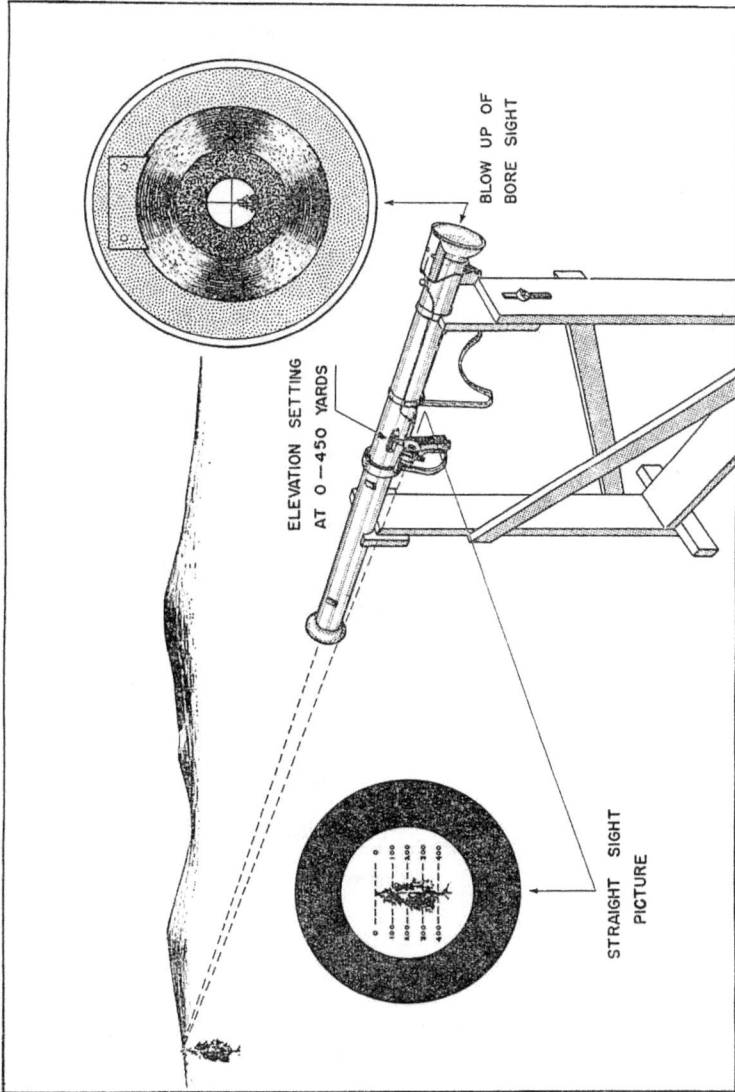

BLOW UP OF
BORE SIGHT

ELEVATION SETTING
AT 0 – 450 YARDS

STRAIGHT SIGHT
PICTURE

Figure 43. Boresighting.

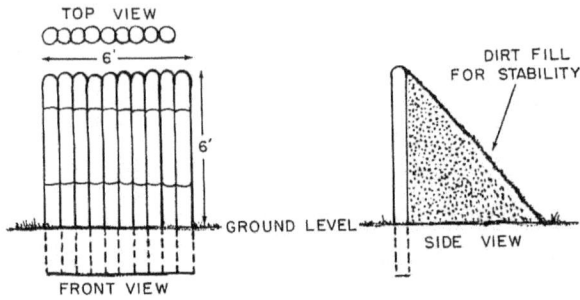

TOP VIEW

6'

DIRT FILL
FOR STABILITY

6'

GROUND LEVEL

SIDE VIEW

FRONT VIEW

LOGS 8:TO 10 IN DIAMETER

Figure 44. Log targets.

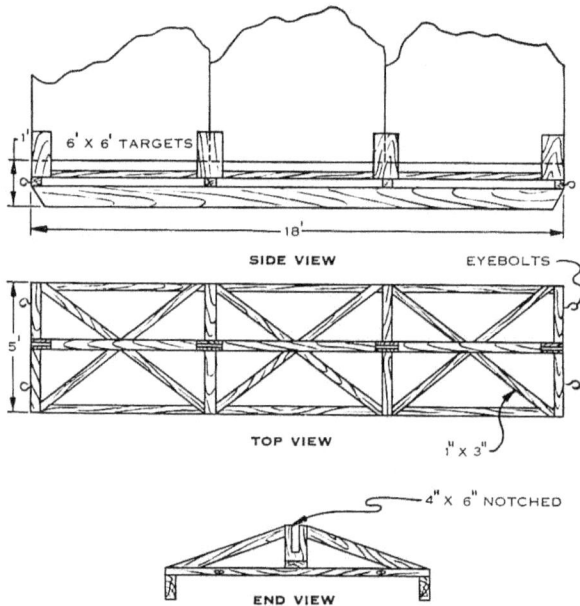

1'

6' X 6' TARGETS

18'

SIDE VIEW   EYEBOLTS

5'

TOP VIEW   1" X 3"

4" X 6" NOTCHED

END VIEW

ALL CONSTRUCTION 2" X 4" EXCEPT AS NOTED

Figure 45. Construction of sled.

the prescribed speed before making the target run. This
distance may be increased when necessary. See figures 46
and 47.

(5) The launcher is loaded before each moving target begins its
run. Practice ammunition is used for firing.

e. Scoring. To be scored as a hit, the target must be struck before
the rocket touches the ground and only while the target is between the

1. TARGET RUNS. TABLE II.

    A. KNEELING: 100-YARD RANGE: 15 MPH: ROUND TO BE FIRED WHILE TARGET MOVES FROM STAKE B TO E.

    B. STANDING: 100- TO 150-YARD RANGE: 10 MPH: ROUND TO BE FIRED WHILE TARGET MOVES FROM STAKE F TO A.

    C. KNEELING: 100- TO 150-YARD RANGE: 10 MPH: ROUND TO BE FIRED WHILE TARGET MOVES FROM STAKE A TO F.

    D. STANDING: 150-YARD RANGE: 10 MPH: ROUND TO BE FIRED WHILE TARGET MOVES FROM STAKE F TO A.

    E. KNEELING: 200-YARD RANGE: 10 MPH: ROUND TO BE FIRED WHILE TARGET MOVES FROM STAKE C TO D.

    2. SLED IS IN POSITION TO BE PULLED BY TRUCK FOR TARGET RUN 1 FROM STAKE D TO STAKE C. TRUCK THEN PULLS OTHER END OF CABLE FOR TARGET RUN 2, ETC. SLED SHOULD BE PROTECTED BY 1-FOOT BANK OF EARTH.

    3. TARGET IS NEVER CONCEALED FROM THE GUNNER. HE MAY BEGIN TRACKING THE TARGET AT THE MOMENT IT LEAVES THE STARTING POINT.

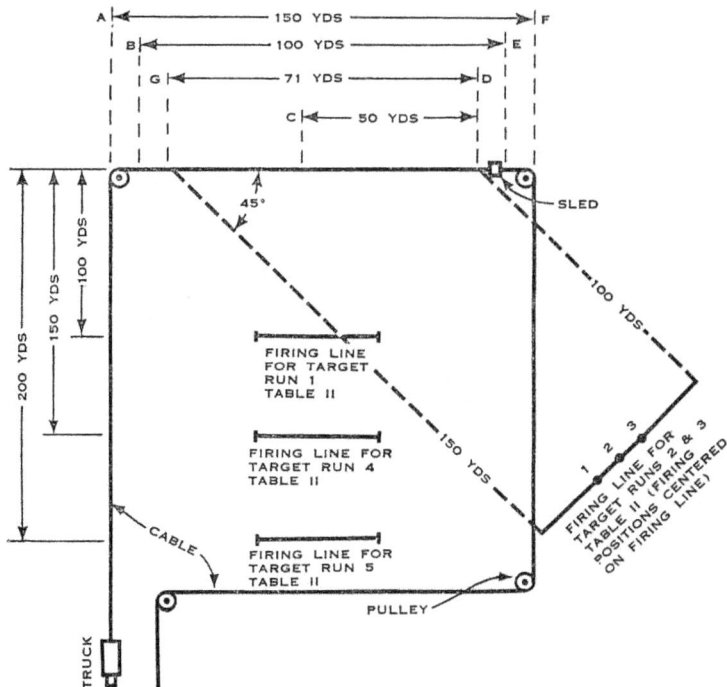

Figure 46. Alternate moving target range.

proper markers. A ricochet hit or a hit at the improper time is scored as a miss.

f. Misfire. When a gunner signifies a misfire (observing procedure outlined in par. 39), the scorer checks the position of the safety, operates the trigger, and attempts to fire. If the rocket fires, he scores a miss for the target. In case of a misfire because of a faulty rocket or a malfunction of the launcher, the gunner is given another rocket to fire.

NOTES

1. TARGET RUNS. TABLE II.

A. KNEELING: 100-YARD RANGE: 15 MPH: ROUND TO BE FIRED WHILE TARGET MOVES FROM STAKE B TO E.

B. STANDING: 100- TO 150-YARD RANGE: 10 MPH: ROUND TO BE FIRED WHILE TARGET MOVES FROM STAKE D TO G.

C. KNEELING: 150- TO 100-YARD RANGE: 10 MPH: ROUND TO BE FIRED WHILE TARGET MOVES FROM STAKE G TO D.

D. STANDING: 150-YARD RANGE: 10 MPH: ROUND TO BE FIRED WHILE TARGET MOVES FROM STAKE A TO E.

E. KNEELING: 200-YARD RANGE: 10 MPH: ROUND TO BE FIRED WHILE TARGET MOVES FROM STAKE F TO A.

2. TARGET IS NEVER CONCEALED FROM GUNNER. HE MAY BEGIN TRACKING THE TARGET AT THE MOMENT IT LEAVES THE STARTING POINT.

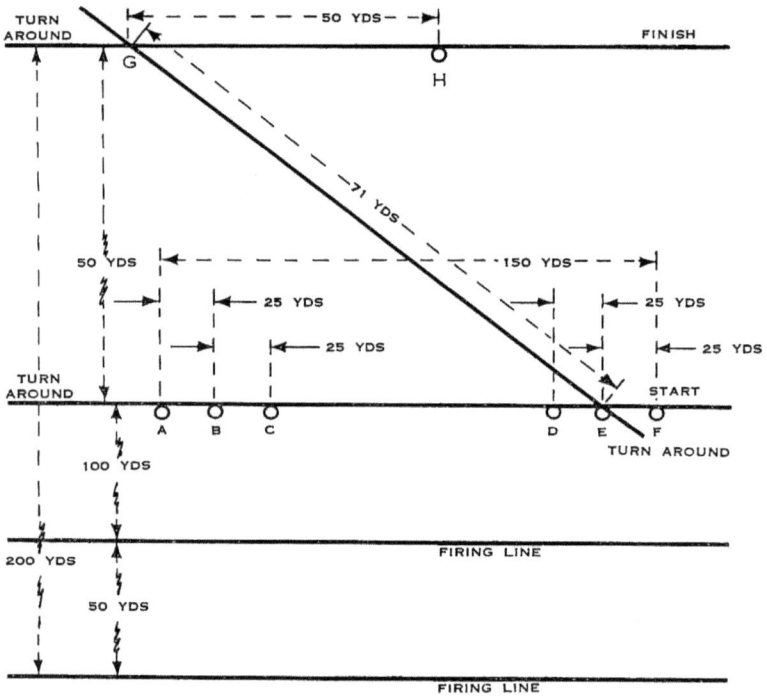

Figure 47. Moving target range.

55. Familiarization Firing

If the gunner is not required to fire the qualification course, he fires the familiarization course. This course is fired with the practice rocket, according to table III.

Table III. Familiarization Course

Range	No. of rockets	Targets	Position
150 yds_____	1	Dummy tank (stationary)_____	Prone.
150 yds_____	2	Dummy tank (stationary)_____	Kneeling.

CHAPTER 8

TECHNIQUE OF FIRE

56. General

Training in technique of fire teaches the rocket launcher team how to place effective fire on a target. It is scheduled after preparatory marksmanship training and range practice. The training includes range estimation, speed estimation, target designation, fire control, fire commands, target vulnerability, and field target firing.

57. Range Estimation

When practicing range estimation, apply only the maximum range of the rocket. Do not practice range estimation on targets moving at excessive speeds. Practice on ranges less than 300 yards (275 meters). (Point targets—50 to 300 yards or 46 to 275 meters.) In making estimates, emphasize speed with accuracy. Practice estimating the ranges to moving targets of different sizes. Practice sight setting at the same time range estimation is practiced. For principles and procedures of range estimation, see FM 23–5.

58. Speed Estimation

a. The 3.5-inch rocket is used primarily against tanks and enemy armored vehicles. Normally, this type of target is moving; therefore, to get a hit, estimate the target's speed, compute the lead, and then apply that lead to the sight.

b. The ability to estimate speed is developed in the same manner as the ability to estimate range—by frequent practice. First, practice speed estimation at known distances, directions, and speeds; but vary the type of vehicle. Then vary the range and vehicle. Finally, vary all four factors and, at the same time, practice sight setting and lead calculating.

59. Target Designation

See FM 23–5.

60. Fire Control

The fire control part of a fire command tells when to fire, when to shift fire, and how many rockets to fire.

61. Fire Commands

a. General. The fire command for the rocket launcher team is the same as the fire commands for other infantry crew-served weapons. It is given as briefly, simply, and completely as possible. Subsequent fire commands are used for adjusting or shifting the fire.

b. Elements of a Fire Command. The elements are given in the following sequence: direction, description of target, range, distribution (for area type fire), and commence fire or cease fire.

(1) The well-trained gunner is the best qualified team member to determine the range, number of leads, and the fire control elements to be used. In most situations, the fire command issued to him consists only of the direction and description of the target. He is given subsequent fire commands to adjust fire on stationary or area targets and, in some cases, on moving targets.

(2) When firing at the stationary and area targets, give range changes in yards and deflection changes in mils. If the range involved is 450 yards or less, use the appropriate range marking on the sight reticle (hold-off may be used for changes less than 50 yards). If the range exceeds 450 yards, move the indicator arm to the appropriate marking on the range scale plate and use the zero line of the sight reticle. Make the mil changes in deflection by using the lead markings of the sight reticle. Each lead line, or space between lead lines, represents 25 mils. For example, to make a deflection correction of "right 25 mils," use as an aiming point the first lead left of the old aiming point.

(3) Normally, for moving targets, the gunner makes his own adjustments based on his observation of the strike of the rocket. However, when a subsequent fire command is given, the range corrections are given in yards and the deflection correction in leads.

(4) Load the weapon on the first element of the fire command.

c. Examples of Fire Commands for Moving Targets.

Initial Fire Command

    Target designation

    RIGHT FRONT

    TANK

    200

In this command, the range is given to designate the target. The range is omitted when there is only one target or when there is only one target in the given direction. In this instance, the gunner estimates his own lead and opens fire when the launcher is loaded.

Subsequent Fire Command
  Fire Control
  CEASE FIRING
Initial Fire Command
  Target designation
  FRONT
  TANK
  150
Fire Control
  TWO LEADS
  COMMENCE FIRING
Subsequent Fire Command (for adjustment)
  DROP 50
  1½ LEADS
Subsequent Fire Command (for shifting fire)
  Target designation
  RIGHT FLANK
  TANK

d. Examples of Fire Command for a Stationary Target.
Initial Fire Command
  Target designation
  FRONT
  PILLBOX
  200
Fire Control
  THREE ROUNDS
  COMMENCE FIRING
Subsequent Fire Command (for adjustment)
  DROP 50
  RIGHT 25 MILS (one lead)
  COMMENCE FIRING
  CEASE FIRING

62. Vulnerability of Armor

a. A tank, without the protection afforded by dismounted infantry, is vulnerable to a close-in attack by well-armed foot troops. When buttoned up, the visibility of the tank crewmen is rather restricted. Tank killer teams should take maximum advantages of this relative "blindness" of enemy tank crews.

b. The lightly armored areas of the tank hull are usually in the rear and on the sides. A gunner should, by maneuver or by waiting for the tank to change position, seek to secure a hit on the tank's most vulnerable part. A rocket hit on either of the tracks will probably immobilize the tank, thus presenting a stationary target which is extremely vulnerable to destruction.

63. Field Target Firing

Field target firing tests the ability of the rocket team to engage a moving target, stationary target, or area type target. For antitank firing practice, the targets are of different types and are towed at various ranges, directions, and speeds. However, they are within effective range of the rocket and at usual battlefield speeds. The objective is to achieve first rocket hits.

APPENDIX I

REFERENCES

JCS Pub 1	Dictionary of United States Military Terms for Joint Usage.
AR 320–5	Dictionary of United States Army Terms.
AR 370–5	Qualification and Familiarization.
AR 385–63	Regulations for Firing Ammunition for Training, Target Practice, and Combat.
FM 21–5	Military Training.
FM 23–5	U.S. Rifle, Caliber .30, M1.
TB 9–2002–1	3.5-Inch Rocket Launcher, M20A1 and M20A1B1.
TM 9–2002	3.5-Inch Rocket Launcher, M20 and M20B1.
TM 9–1950	Rockets.
TM 9–247	Ordnance Maintenance: Materials Used for Cleaning, Preserving, Abrading, and Sealing Ordnance Material and Related Materials Including Chemicals.
TM 9–1900	Ammunition, General.
DA Pam 108–1	Index of Army Motion Pictures, Film Strips, Slides, and Phono-Recordings.
DA Pam 310–1	Index of Administrative Publications.
DA Pam 310–3	Index of Training Publications.
DA Pam 310–4	Index of Technical Manuals, Technical Bulletins, Supply Bulletins, Lubrication Orders, and Modification Work Orders.
DA Pam 310–5	Index of Graphic Training Aids and Devices.

APPENDIX II

DESTRUCTION OF ROCKET LAUNCHER AND ROCKETS

1. General

a. When capture of the rocket launcher is imminent, or if it becomes necessary to abandon the weapon in the combat zone, the unit commander, in accordance with instructions from higher command, orders its destruction.

b. Adequate destruction requires all parts essential to the operation of the weapon be destroyed or at least damaged beyond repair. Priority is given to the destruction of those parts most difficult to replace, such as the sight and firing mechanism. *It is important that the same parts of all launchers be destroyed so the enemy cannot construct one complete unit from several damaged ones.*

2. Method of Destroying Launcher

a. Disassemble the launcher.

b. Using a heavy implement, dent and deform the barrel and stock.

c. Smash the sight, firing mechanism, barrel coupling lock, barrel latch, and contactor latch.

d. Slash the sling.

e. Scatter the damaged parts over a wide area, or place them with other materials to be destroyed by fire or explosive.

3. Destruction of Rockets

a. Rockets are best destroyed by burning. Stack packed or unpacked rockets in piles so the rocket heads point toward the enemy, or place them in a trench with their heads downward. Place paper, rags, brush, or wood around and on the pile. Pour gasoline and oil over the pile. Use flammable material to insure a fire hot enough to destroy the rockets. Ignite and take cover.

b. Ignition of the rocket propellant will cause some rockets to be projected in unpredictable flight. Therefore, the danger area for the destruction of rockets should have a radius equivalent to the effective range. Generally, however, the rocket will travel in the same direction in which it is pointed when ignited.

APPENDIX III

LAUNCHERS, M20 AND M20B1

1. Difference Between Models

a. The difference between the M20 and M20B1 launchers is in the fabrication of the front and rear barrels (figs. 48 and 49). The tubes of the M20 launcher are manufactured from aluminum tube stock. The component parts of the barrel assembly, such as the breech guard and the barrel coupling screw and nut, are fastened to the tube by means of screws. The tubes of the M20B1 launcher are aluminum castings. Many of the component parts of the barrel assembly are cast as part of the tube, resulting in a slight reduction in overall weight.

b. The difference between the M20 and M20B1 launchers and the M20A1 and M20A1B1 launchers (fig. 50) is the addition of the contactor latch group assembly on the latter. This replaces the contactor latch housing on the M20 and M20B1 launcher.

Figure 48. Front and rear barrels of launcher, M20—left side view.

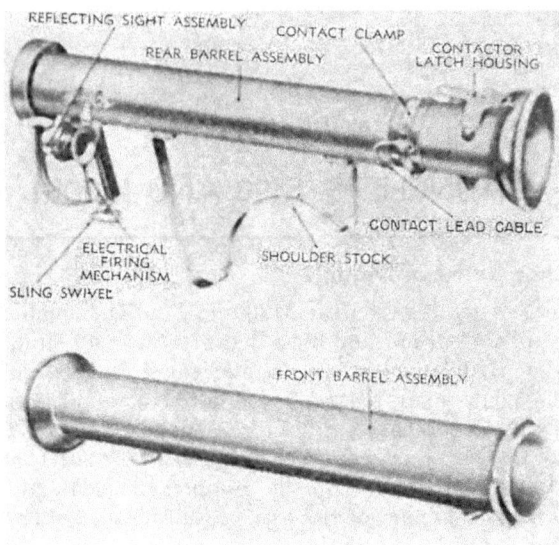

Figure 49. Front and rear barrels of launcher, M20B1—left side view.

2. Tabulated Data for Launchers

Length of launcher assembled (for firing) (approx)_____ 60 inches
Weight of M20 launcher with bipod and monopod
(approx)_____ 15 pounds
Weight of M20B1 launcher with bipod and monopod
(approx)_____ 14 pounds
Length of front barrel_____ 30 inches
Weight of M20 front barrel with bipod (approx)_____ 6 pounds
Weight of M20B1 front barrel with bipod (approx)_____ 6 pounds
Length of rear barrel (approx)_____ 31 inches
Weight of M20B1 rear barrel with monopod (approx)__ 8 pounds
Weight of M20 rear barrel with monopod (approx)_____ 9 pounds

3. Launcher Controls and Sighting Equipment

a. General. This paragraph describes, locates, and illustrates the controls and instruments provided for the operation of the M20 and M20B1 rocket launcher. The trigger, safety switch, barrel coupling lock lever, barrel latch handle, sighting equipment, breech guard, stock, sling, and muzzle flash deflector are the same as on the M20A1 and M20A1B1 launchers.

b. Contactor Latch Housing.

(1) On the M20 launcher, the contactor latch housing is mounted on the contactor latch clamp in front of the breech guard (fig. 48). On the M20B1 launcher, the contactor latch housing is mounted in front of the breech guard on lugs

68

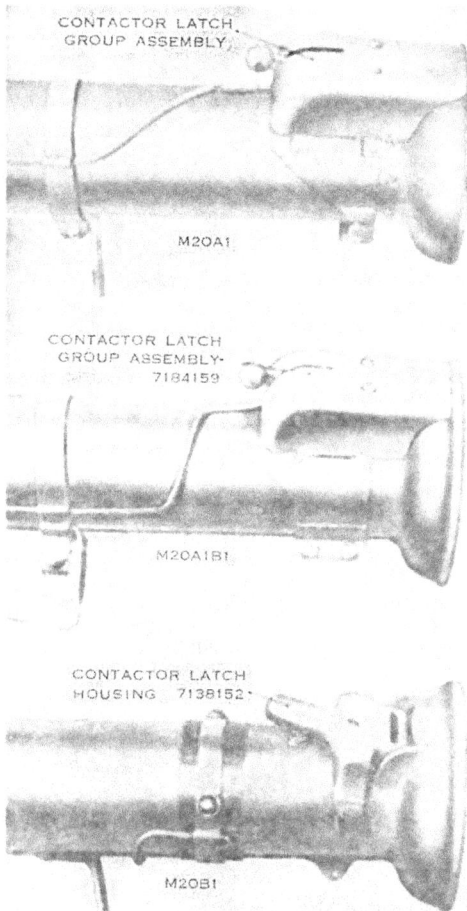

Figure 50. Comparison of breech end of 3.5-inch rocket launchers, M20A1, M20A1B1, and M20B1.

cast as a part of the barrel (fig. 49). The latch housing protrudes through an opening in the breech guard.

(2) A steel blade, riveted to the rear of the body of the contactor latch housing, engages in the unpainted groove at the rear of the fin assembly. This holds the rocket in its proper firing position in the launcher. The contact between the blade of the contactor latch housing and the unpainted groove on the rocket fins ground the rocket to the rear barrel, completing the firing circuit. A spring keeps the rear end of the contactor latch housing depressed.

(3) When the forward end of the contactor latch housing is pressed down against the barrel, the blade at the rear of the

contactor latch housing swings clear of the breech of the launcher tube. This is done during loading or unloading of the launcher.

   c. Contact Lead Cable, Contact Spring Clamp, and Contact Springs.

      (1) The contact lead cable (figs. 48 and 49) is an insulated wire which connects the firing mechanism in the trigger grip to the contact spring clamp.

      (2) The contact springs are coiled springs mounted on the sides of the contact spring clamp (fig. 51). Electrical contact is established only when the trigger is squeezed.

4. Operation and Functioning

   a. Front and Rear Barrel Groups.

      (1) See paragraph 23.

      (2) The contact springs (figs. 51 and 52) on the M20 and M20B1 launchers complete the firing circuit of the launcher when

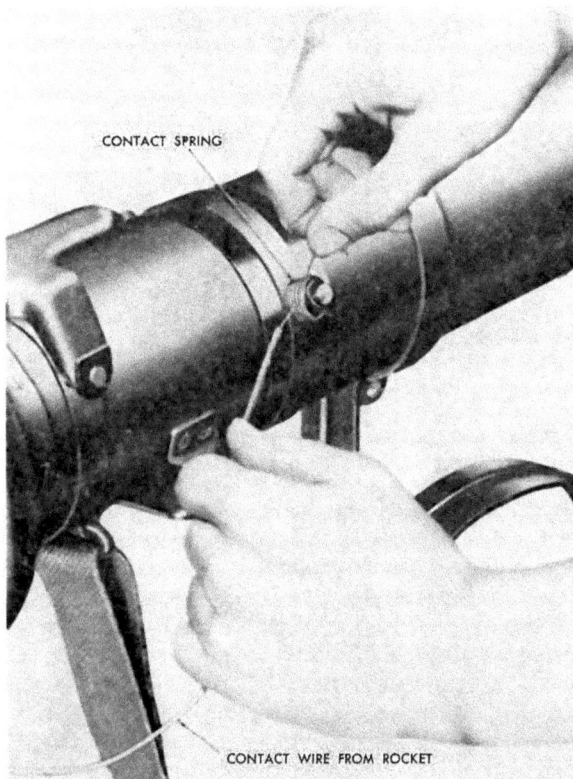

CONTACT SPRING

CONTACT WIRE FROM ROCKET

Figure 51. Placing contact wire to contact spring.

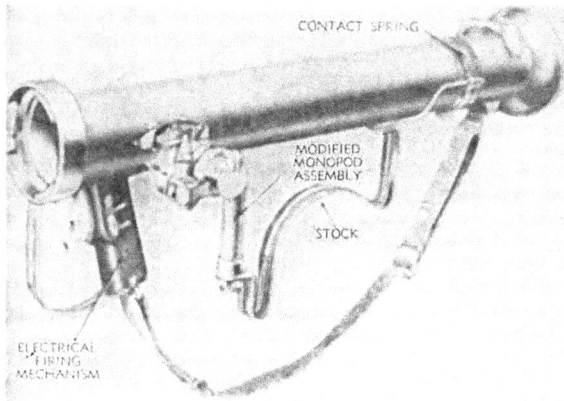

Figure 52. Rear barrel of launcher, M20B1—left side view.

the blue contact wire of the rocket is forced between the coils. The contact spring clamp mounts both springs and connects them together electrically so either contact spring may be used. There is an insulating band around the barrel under the contact clamp.

(3) The contact lead cable, encased in an aluminum tube, connects (fig. 48) the contact clamp to the firing mechanism. An insulator sleeve on the end of the aluminum tube separates the contact clamp from the tube. The firing mechanism is grounded to the rear barrel.

b. Functioning.

(1) When a rocket has been loaded into the launcher and the long contact wire of the rocket has been engaged in a contact spring, the electrical circuit is complete. When the trigger is squeezed, the magneto generates current. The current leaves the trigger mechanism and moves through the contact lead cable to the contact spring clamp. The contact spring clamp, in turn, carries the current to the contact springs. This part of the circuit is insulated from the barrel.

(2) The long contact wire of the rocket picks up the current at the contact spring and carries it to the electrical igniter in the rocket. To complete the circuit, the short contact wire in the rocket conducts the current from the igniter to the aluminum support ring. The blade of the contactor latch, engaged in the unpainted groove of the support ring, picks up the current. The current then passes through the contactor latch housing, through the barrel, and back to the trigger mechanism.

5. Ammunition

The ammunition currently available can be fired from all models of the rocket launchers. For additional information, see chapter 4.

6. Maintenance

a. *Disassembly and Assembly of Launcher.* See chapter 6.

b. *Operational Inspection.*

   (1) *Purpose.* See paragraph 41.

   (2) *Procedure.*

      (a) Examine the contact springs to see they are soldered securely to the clamp and are clean and free of rust, paint, and grease.

      (b) See paragraph 42 for additional procedures.

c. *Preventive Maintenance.* See paragraph 43.

d. *Repairs.*

   (1) Replace the contactor latch spring when the contactor latch does not function.

   (2) Clean the corroded electric contact points with a crocus cloth.

   (3) Additional repairs, see paragraph 44.

e. *Lubrication Order.* See paragraph 45.

   (1) *Contactor latch pins.* Apply one or two drops of preservative lubricating oil to the contactor latch pins weekly, monthly, and after firing.

   (2) *Contactor latch housing.* Clean at intervals specified in the lubrication order. Remove any powder fouling or rust with rifle bore cleaner, drycleaning solvent, or volatile mineral spirits. Crocus cloth may be used to remove deep seated corrosion. Wipe dry and protect with a film of lubricant.

   (3) *Additional information.* See paragraph 46.

f. *Operation Under Unusual Conditions.* See paragraph 47.

7. Preparatory Marksmanship Training

a. *Purpose.* See paragraph 31.

b. *Aiming.* See paragraph 32.

c. *Trigger Squeeze.* See paragraph 33.

d. *Positions.* See paragraph 34 through 37.

e. *Loading Launcher.*

   (1) For the gunner's duties, see paragraphs 38 and 39.

   (2) The loader (also see pars. 38 and 39)—

      (a) Grasps the rocket by the motor tube and pulls the blue lead wire (fig. 53) out of the expansion cone at the rear of the rocket. He straightens the end of the wire and pulls off the insulating tube.

      (b) Removes the safety band from the rocket.

Figure 53. Pull out long (blue) contact wire.

(c) Grasps the rocket with the ejection pin pointed down or to the side. In this position, the pin will not strike the contactor latch blade.

(d) Grasps the rear of the launcher with his right hand, palm down. He places his thumb under the barrel and his fingers on the forward end of the contactor latch.

(e) Presses down with his fingers and raises the blade of the contactor latch to clear the way for the rocket. He inserts the rocket into the breech of the launcher.

(f) Moves his left hand to the fin assembly but not over the opening of the rocket motor. He slides the rocket forward into the launcher until the blade of the contactor latch, when released, fits into the unpainted groove of the rocket fin assembly (fig. 54). To insure effective electrical contact, he gives the rocket a slight rotary twist with his left hand.

(g) Places the stripped end of the long contact wire in either of the two contact springs (fig. 55).

Figure 54. Slide rocket into launcher and press the contactor latch.

 (*h*) Checks to see no part of the uninsulated portion of the contact wire touches any metal before touching the contact spring.

 (*i*) Glances to the rear to see that the backblast area is clear.

 (*j*) Taps the gunner and calls UP.

 f. Malfunction and Immediate Action.

  (1) *Malfunctions.* Malfunction of the launcher is defined as a failure to function satisfactorily. Some of the more common malfunctions and the corrective measures used to reduce them in the field are as follows:

  (*a*) *Failure to load.*

Cause	Correction
Bent or defective contactor latch____	Attempt to straighten the latch. When contact cannot be made, replace the latch.

(b) *Failure to fire.*

Contact wire not securely connected to the contact spring.
    Attach securely.

No contact between the contactor latch and the unpainted groove of the rocket fin assembly.
    Rotate the rocket and, at the same time, press down on the rear of the contactor latch to seat the blade of the contactor latch firmly into the unpainted groove.

Figure 55. Attach long contact wire to contact spring

(2) *Immediate action.* Immediate action is the prompt action taken by the firer to reduce a stoppage. If a misfire occurs, perform immediate action as follows:

(a) For the gunner's initial duties, see paragraph 39.

(b) The loader—

1. Repeats MISFIRE.
2. Waits 15 seconds (counting slowly to 15) to allow for a possible hangfire.
3. Checks the long contact wire to see that the uninsulated portion does not touch the launcher at any point other than at the contact spring.
4. Disengages the long contact wire and presses down on the rear of the contactor latch.
5. At the same time, rotates the rocket to insure electrical contact between the blade of the contactor latch and the unpainted groove of the fin assembly.
6. Engages the long contact wire and checks to see it is clear of the launcher.
7. Checks the backblast area and taps the gunner, calling UP.

(c) For the gunner's duties, see paragraph 39.

(d) The loader—

1. Repeats UNLOAD.
2. Waits 15 seconds, counting slowly to 15.
3. Removes the long contact wire and withdraws the rocket from the launcher.
4. Replaces the safety band over the borerider and the shorting clip on the copper band.
5. Puts the rocket aside for repacking and disposal.

g. *Tracking Moving Targets.* See paragraph 40.

INDEX

G. H. DECKER,
General, United States Army,
Chief of Staff.

Official:

J. C. LAMBERT,
Major General, United States Army,
The Adjutant General.

Distribution:
  Active Army:

DCSOPS (2)	USA Sig Tng Comd (5)
ACSRC (2)	USATTC (5)
TPMG (1)	MDW (5)
CofOrd (1)	Armies (5)
CofEngrs (1)	Corps (3)
TQMG (1)	Div (2) (ea CC (1))
CofT (1)	Regt/Gp/BG (1) except Inf BG
CSigO (1)	(5)
CCmlO (1)	Bn (5)
USCONARC (5)	Co/Btry (5)
USAIB (3)	USAIS (1918)
USA Arty Tng Comd (5)	USAARMS (200)
USA Cmbt Survl Tng Comd (5)	USASCS (5)
USAQMTC (5)	USA Ord Sch (40)
CMLCTNGCOM (5)	USAES (10)
USA Ord Tng Comd (5)	USACMLCSCH (5)

NG: State AG (3); units—same as Active Army except allowance is one
    copy to each unit.
USAR: Same as Active Army except allowance is one copy to each unit.
For explanation of abbreviations used, see AR 320–50.

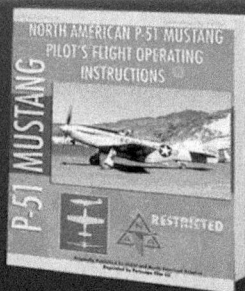

www.ingramcontent.com/pod-product-compliance
Lightning Source LLC
Chambersburg PA
CBHW052206090426
42741CB00010B/2425